Between Inner Space and Outer Space

John D. Barrow is Professor of Mathematical Sciences in the Department of Applied Mathematics and Theoretical Physics at Cambridge University, and Director of the Millenium Mathematics Project, a new programme to raise the profile of mathematics amongst young people and the general public. His principal area of scientific research is cosmology, and he is the author of several highly acclaimed books about the nature and significance of modern developments in physics, astronomy, and mathematics, including *The Left Hand of Creation*, *The Anthropic Cosmological Principle*, *The World Within the World* (a revised edition of which, entitled *The Universe that Discovered Itself*, is to be published in 2000), *Theories of Everything*, *Pi in the Sky*, *The Origin of the Universe*, *The Artful Universe*, and *Impossibility*.

Between Inner Space and Outer Space

Essays on Science, Art, and Philosophy

JOHN D. BARROW

Centre for Mathematical Sciences
University of Cambridge

OXFORD
UNIVERSITY PRESS
Great Clarendon Street, Oxford OX2 6DP
Oxford University Press is a department of the University of Oxford
and furthers the University's aim of excellence in research, scholarship,
and education by publishing worldwide in
Oxford New York
Athens Auckland Bangkok Bogotá Buenos Aires Calcutta
Cape Town Chennai Dar es Salaam Delhi Florence Hong Kong Istanbul
Karachi Kuala Lumpur Madrid Melbourne Mexico City Mumbai
Nairobi Paris São Paulo Singapore Taipei Tokyo Toronto Warsaw
with associated companies in Berlin Ibadan
Oxford is a registered trade mark of Oxford University Press
in the UK and in certain other countries
Published in the United States
by Oxford University Press Inc., New York

First published 1999
First issued as an Oxford University Press paperback 2000

British Library Cataloguing in Publication Data
Data Available
Library of Congress Cataloging in Publication Data
Data Available
ISBN 0-19-288041-1
1 3 5 7 9 10 8 6 4 2
Typeset by
Footnote Graphics, Warminster, Wilts
Printed in Great Britain by
Cox & Wyman Ltd,
Reading, Berkshire

In memory of my mother

Preface

Public interest in science has grown steadily since the late 1970s. Specialist science journalists can be found on the staffs of all serious newspapers; popular science is the biggest non-fiction section in most bookstores; scientists are commonly to be found on screen and radio describing their special interests and injecting their insights into the wider discussion of ideas. Ever since I was a postgraduate student I have written reviews and explanatory articles for newspapers and magazines, both in the English-speaking world and in Europe. In this book some of the articles have been organised to provide a series of gentle introductions to key problems in astronomy, physics, and mathematics. The articles vary in length and style and fall into ten sections, each introduced by an essay outlining the issues under discussion and providing the context in which the original articles were written. The subjects chosen reflect some of the important developments that have taken place on the frontiers of fundamental physics and astronomy over the last twenty years. They encompass the advances in our thinking about the early history of the Universe, the utility of mathematics, time, complexity and chaos, the quest for an ultimate Theory of Everything, our Universe (and others) as a habitat for life, quantum reality, and the inter-relationships between science, religion and aesthetics.

I am grateful to Oxford University Press for their invitation to produce this compilation and especially to Martin Baum, John Grandidge, and Susan Harrison for their enthusiasm and editorial assistance. Ingram Pinn provided the illustrations in his inimitable style. I would also like to thank all those who assisted with the preparation of the original articles for publication and those collaborators and colleagues whose research, books, and discussions were stimuli for so many of them. Elizabeth helped from the beginning, providing life, simplicity, and complexity in measure where needed, while our children, David, Roger, and Louise, were surprised that anyone wrote essays that they didn't have to hand in on Monday morning.

Brighton JDB
October 1998

Acknowledgements

Many of the articles in this volume have appeared previously and the following sources are gratefully acknowledged for permission to reproduce them.

Chapter 1: *The Daily Telegraph*, 15 November 1995.

Chapter 2: *The Times Higher Education Supplement*, 16 April 1993.

Chapter 3: *The Times Higher Education Supplement*, 25 May 1990.

Chapter 4: From *Vistas in Astronomy*, 37, 409–427, 1993.

Chapter 5: Based on a *Focus* magazine article, October 1993.

Chapter 6: *The Times Higher Education Supplement*, 14 January 1983.

Chapter 7: *The Guardian*, 13 October 1994.

Chapter 8: *The Times Higher Education Supplement*, 9 July 1993.

Chapter 9: *Nature's Imagination*, OUP, 1995.

Chapter 10: *Boundaries and Barriers*, ed. J. Casti and A. Karlqvist, Addison Wesley, Mass., 1996.

Chapter 11: *Nature*, **329**, 771–772, 1987.

Chapter 12: *The Ottawa Citizen*, 6 April 1998.

Chapter 13: *Scienza e Technica Italia*, 1990 Yearbook, pp. 10–15.

Chapter 14: *The Observer*, 16 May 1993.

Chapter 15: *The Guardian*, 21 October 1993.

Chapter 16: *The Daily Telegraph*, 1 May 1997.

Chapter 17: *Liberal*, 12 March 1998.

Chapter 18: *New Scientist*, 21 April 1990.

Chapter 19: *New Scientist*, 6 March 1993.

Chapter 20: *Il Caso è la Libertia*, ed. M. Ceruti *et al.*, Laterza, Rome, 1994.

Chapter 21: *The Times Higher Education Supplement*, 31 May 1996.

Chapter 22: From *Theories of Everything*, OUP, 1991.

Chapter 23: *The Daily Telegraph*, 9 December 1995.

Chapter 24: *The Times Higher Education Supplement*, 23 June 1995.

Chapter 25: *The Times Higher Education Supplement*, 15 December 1995.

Chapter 26: *Faces of Time*, ed. E. Mariani, IPS, Naples, 1997.

Chapter 27: *The Observer*, 9 May 1993.

Chapter 28: *Nature*, 19 September 1996.
Chapter 29: *The Times Higher Education Supplement*, 22 August 1980.
Chapter 30: *The London Review of Books*, 20 August 1992.
Chapter 31: *The Times Literary Supplement*, 22–28 April 1988.
Chapter 32: *The Times Literary Supplement*, 10 January 1992.
Chapter 33: *The Times Higher Education Supplement*, 21 June 1991.
Chapter 34: *New Scientist*, 9 February 1990.
Chapter 35: *The Times Higher Education Supplement*, 6 July 1984.
Chapter 36: *Sfera Magazine* **34**, 16–19, 1993.
Chapter 37: *The Independent Magazine*, 2 October 1994.
Chapter 38: *New Scientist*, 3 November 1990.
Chapter 39: University of Sussex, Annual Report 1992/3.
Chapter 40: *Scienza, Filosofia e Teologia di Fronte alla Mascita dell' Universo*. ed. New Press, Como, 1997, pp. 79–94.
Chapter 41: *New Scientist*, 16 March 1996.
Chapter 42: Unpublished.

Contents

Ingram Pinn

Part 1

The popularisation of science

Being popular is important.
Otherwise, people might not like you.
Mimi Pond, *The Valley Girl's Guide to Life*, 1982

The popularisation of science has finally become respectable. Fifty years ago, a Jeans or an Eddington risked condemnation by friends and foes alike for their sallies into the popular press. But today the simplification of science is welcomed on many fronts. Scientists seek to convince the public of the value and excitement of their work, funding agencies seek public and political support for the projects that taxpayers' money is underwriting, and the media seek to meet and extend the public's curiosity to learn about new ideas and unusual personalities. What has changed? Perhaps it is the gradual encroachment of science upon issues that were once the sole preserve of the humanities. Ultimate questions about the origins of life, intelligence, human behaviour, the Universe, and everything else, have ceased to be solely matters of speculative philosophy and theology. Scientists have found new things to say about these problems that are not merely restirrings of the brew of philosophical opinions we have inherited from thinkers of the past. If there is a pattern to successful popular science writing in recent years it has been its engagement with these deep questions. It exploits the fact that it is the only real news and frontiers are the places to look for it. At one end of the size spectrum there is the inner space of the most elementary particles of matter and the perplexing puzzle of space and time itself; at the other end lies the outer space of stars and galaxies which constantly surprise us with the drama of their cataclysmic evolution, so spectacularly captured by the mirror of the Hubble Space Telescope—a glass that enables us to see, albeit darkly, back to the beginnings of the universe around us. In between these extremes we sit, the cosmic middle classes, wondering how it was that events contrived to make a planet where conditions were temperate enough for the evolution of minds almost complex enough to understand themselves.

The popularisation of science has not only been pursued by scientists. A cadre of specialist journalists now fills significant parts of our newspapers and magazines with the latest news about our attempts to unravel the structure of the Universe and its laws. Some branches of the media have gone further still in an effort to stimulate young people to think seriously about the craft of explaining science to the outsider. The first chapter in this section was written in response to an invitation from the *Daily Telegraph* to launch their annual Young Science Writer of the Year competition. It offers some advice to new writers, both with regard to style and how they might decide what to write about. There is much scope to be different: although much has been written about the aforesaid 'meaning of life' issues, we are surrounded by all manner of curious devices and natural events that the average reader knows little or nothing about.

When one gets down to the business of explaining unusual scientific ideas there is one time-honoured device that is repeatedly drawn into play: that of analogy. The popularisation of science in the Anglo-Saxon tradition draws greatly upon this explanatory tool. The presentation of quarks as marbles rattling around in a rubber bag, of superstrings as elastic bands, or of pulsars as lighthouses, has become the essence of the art of explaining. Some reviewers have been known to judge the success of a popular science book by the number of distinctive new analogies that it brings to the exposition of its subject matter. The second chapter, '*In the world's image*', discusses this love of analogy. It is a relic of a particularly 'British' style of doing science at the turn of the last century, particularly associated with Lord Kelvin, which attracted some criticism from those on the continent, who were content with the abstractions of mathematics. It was not necessary to translate them into little pictures of simple machines before saying one had understood the phenomenon under study. This article was partly inspired by the unusual statement that is the crux of Stephen Hawking's successful account of modern attempts to understand the origin of the Universe and the nature of time (*A brief history of time*). The idea that 'time becomes another dimension of space' is disarmingly simple to state; we understand the meaning of every single word; but does it convey any information to the general reader? Is there another way to explain it? Perhaps this is one example of an idea that does not admit of a simple analogy in terms of everyday experience. If so, is that a sign of its profundity or of the poverty of our imaginations?

Finally, '*Falling between two cultures*' is a review of a *pot pourri* of unusual writing about science, not all of it falling under the label of 'popularisation'. There are many such books (this is one of them!) but the collection of Walter Gratzer's contains much that is charmingly unexpected and this review takes the opportunity to throw in some other more off-beat examples for good measure.

CHAPTER 1

As well as science what do you know?

Can you describe a spiral staircase without moving your hands?

A. Mathias

So you want to be a science writer. 'Listen,' someone told me, 'I've been writing for just a few months and I've already sold several articles—my coat, my typewriter, and my watch.' So how do you join this lucrative profession? Anyone can have a go; but asking how to make a success of it is a little like asking the gardener of Trinity College how he gets his lawns to look so immaculate. 'Just roll and water for 600 years sir.'

The moral is that it is a gradual process. There needs to be groundwork: the establishment of a base of accurate information, a grip on the grammar of language, and an ability to apply the seat of your pants to the seat of your chair. But don't lock yourself away in a lighthouse, because you also need an eye for the world around you: an ability to see everyday things in new ways and a familiarity with matters of interest to the public and the press. Keep your eye on a wide range of newspapers and magazines. You will find it difficult to write effectively for publications of which you are not a reader.

Your own grasp of science will determine the limit of your ability to shed light on developments at its frontiers. Let common sense be your guide. Don't attempt a confident pronouncement on the significance of relativistic quantum field theory if you know less about it than the average reader, or their cat. But don't give up. Why not try to interview some experts? Be what you are, a typical reader wanting to find out something about a subject of which you know almost nothing. Ask the questions needed to discover what it's all about. Find out about the individuals who do the research: Why do they do it? How do they work? What else do they do? They might just be looking for someone like you to write a popular article.

Get used to obtaining information from the primary source—the scientists doing the work—not from other popular accounts of it. In the process you will accumulate a collection of useful contacts.

Fortunately, there is more to science writing than writing about science. Your success or failure will hinge largely upon your knowledge of other things. Most scientists make poor literary expositors because they know little beyond their own field and write only for like-minded experts in research journals that make few linguistic demands upon them. Scientists are extreme specialists. By contrast, the successful science writer is more often a divergent thinker. Two of my favourites, Primo Levi (*The periodic table*) and Diane Ackerman (*A natural history of the senses*), could not be called science writers at all, but use science to illuminate the many aspects of the human condition. Another favourite, Alan Lightman (*Einstein's dreams*), informs fiction with science in a unique way. There is something unexpected and oblique about the writing of each of these authors that shows up the average scientist's limited experience of the world and its inhabitants.

Be encouraged: because of this blindspot you may one day be surprised to discover that you have made a better job of writing about Professor Jargonbender's Nobel-prizewinning discovery (and Professor Jargonbender) than ever he could himself.

Good writing encourages good reading: good writing is writing that is read, and then re-read. What you read and re-read will colour your style. Its quality will set the standard you aspire to reach. It should also teach you of the need to adopt different styles for different audiences. Charles Darwin's prose style sounds very impressive in Victorian drawing-rooms but would not impress the editor of a mass circulation newspaper today. Know your audience.

As in most things, practice makes perfect. Look for opportunities to write for others. Letter writing is a much neglected art and book reviewing provides an ephemeral outlet that offers a ready-made answer to that great question of human existence, 'What shall I write about?'

The word processor will probably be the tool of your trade. It is a blessing and a curse. Its positive features are obvious, but it will make your prose baggy and indirect; it tempts you to reuse sentences and paragraphs, to move them around, in fact, to do anything except delete them. Revision should be a process of simplification: removing redundant words, dividing long sentences, choosing the best words, and placing them in the best possible order.

A little humour helps make science more digestible. Can you make good use of it? Some parts of your article are of special importance. The title is vital. Does your opening sentence capture the reader's attention? In your final sentence you should have the reader's total attention. Make the most of it. Does what you write pass the test of the tongue—can it be read aloud? Try it. Does it sound right, does it flow, does it have continuity, does it make sense? Generally, good speakers make good writers. It is no accident that the greatest writings in the English language, the King James Bible and Shakespeare's plays, were written to be read aloud.

If you can't explain science in simple words to your friends at the pub you probably won't be able to do it on paper either. Verbal explanation also brings you into direct contact with listeners and typical readers. You learn immediately what they find hard to understand, where your attempts at explanation flounder, and what questions they want answering. Remember the three golden rules: think no jargon, speak no jargon, write no jargon.

Okay, so you've got style, you know a little science, and you know how to check your facts; but how do you find something to write about? Your greatest gift is curiosity. Encourage it; it will help you to identify those everyday phenomena that others overlook. Interesting science is all around you. Develop an eye for the unexpected connections between things, or for those unusually simple questions that children ask: Why is the sky dark at night? Why is the sky blue? Why are there only two sexes? How big could a tree grow? How does a photocopier work? Why don't we fall through the floor?

Choose your subject carefully. Be original. The popularisation of evolutionary biology, quantum physics, cosmology, and consciousness is pretty well worked over. Why not be different? The modern world surrounds us with machines about whose workings most people have little or no knowledge. Why not enlighten them? Most of the population take an interest in sport without noticing its science: the flight of the frisbee, the spin and swing of a cricket ball, the swerve of footballs and golfballs.

The Olympic Games are coming up; have you ever wondered why high-jumpers use the Fosbury Flop, which weight-lifter is pound-for-pound (sorry kilogram-for-kilogram) the strongest, or how ice-skaters manage to spin so fast? Closer to home, how many people know how their television works or why a hot iron removes the creases from their clothes? On to the kitchen, where you could develop a taste for the chemistry of cooking. Next stop, try the bathroom, where the chemical mysteries of perfumes, lathers, and soaps, together with the acoustics of singing in the shower and the meteorology of condensation, should keep you busy for a while. Still stuck? Out to the garden for the subtleties of the greenhouse and the symmetries of snowflakes and icicles. After all this you're going to need a stiff drink. But have you ever wondered why wine is so .... Sorry, now it's your turn.

CHAPTER 2

In the world's image

I never satisfy myself until I can make a mechanical model of a thing. If I can make a mechanical model, I understand it.

Lord Kelvin

The popularisation of science has become a genre of our times: a bridge between the incomprehensible world of science and the comfortable familiarity of the arts. Its history is long. Not least among the reasons for the displeasure breathed upon Galileo was his habit of divulging his revolutionary discoveries to the educated general public in the vernacular. In Newton's time popularisation in England began to have an ulterior motive. The unveiling of the laws of Nature to reveal the universality, simplicity and harmony behind the disorder of appearances was presented as the ultimate proof of the design of Nature and hence of a grand designer behind the scenes. While few had read Newton, many had read about him and there sprang up a crop of eccentric popular books presenting 'Newtonian' theories of just about everything, like Desagulier's *Newtonian system of the world, the best model of government*. The most successful was probably Count Francesco Algarotti's *Sir Isaac Newton's philosophy for the use of the ladies*, which was not quite as condescending as it sounds. It was something of an enlightened novelty to present physics and mathematics to such an audience.

Recently, scientists in the UK have been under attack from commentators who see the exposition of new ideas in science, rather than simply the reporting of new discoveries, as undermining the fabric of society and human values. They have argued that modern science creates an environment in which certainty is indefensible, conservatism is eroded and humanity irrevocably diminished. In response, it is argued that the Universe is not constructed for our convenience. We just have to make the best of it. Humanity displays its moral and ethical qualities by the responses it makes to the world, not by suppressing the truth about it because it might be socially unhealthy. The traditional guardians of the debate over the 'meaning of life' seem to feel threatened by the entry of new scientific concepts into their domain. Faced with being excluded by lack of familiarity with new ideas and issues, a minority have over-reacted against the entire scientific enterprise. The fan of science and its popularisation, on the other hand, sees it as the only real news. Everything else is a variation on a familiar theme of human relationships, politics, war and peace. The danger here is in getting carried away with the immediate consequences of new discoveries so their importance is exaggerated and everything is ultimately devalued. One has only to review the publicity surrounding the discovery of temperature variations in the cosmic background radiation by the COBE satellite, and the ensuing backlash, to appreciate this point. Scientific knowledge should become part of everybody's general knowledge, not merely the fashionable flavour of the month.

Scientific popularisation in English has inherited an interesting tradition, at least in its efforts to communicate progress in the harder physical sciences. If we sample a typical account of the frontiers of particle physics, or of astrophysics, we repeatedly encounter a familiar pedagogic device: the use of analogy. 'The confinement of quarks is like marbles inside a rubber bag'; 'superstrings are like elastic bands'; 'the quantum wave function is like a crime wave'; 'a pulsar is like a lighthouse'; and so on. This approach to the popularisation of science has a distinctly Victorian flavour to it. It reduces the world to a collection of images of practical devices familiar from everyday life. And that giant of Victorian science, Lord Kelvin, made constant use of this approach not only in his explanatory writings but in his research work as well. Interestingly, there were objectors to this style of thinking. Some continental mathematical physicists found it exasperating that British physicists like Kelvin were never content with a mathematical explanation or theory of any natural phenomenon until they could translate it into an analogical picture involving little wheels or rolling balls. While they were content with the abstract mathematical explanation, a Kelvin or a Faraday was not.

An interesting sidelight upon this state of affairs is provided by Stephen Hawking's staggeringly successful book *A brief history of time*. There is one area of this book which seems to stump many general readers and ought to stump experts, only they are so used to it that they pass hurriedly by unperturbed. The point in question is the radical hypothesis proposed by Hawking and Jim Hartle for the initial quantum state of the Universe. A key ingredient of their proposal is the requirement that time becomes another dimension of space in the quantum gravitational environment provided in the first moments of the Big Bang. This has all manner of interesting ramifications, some of which the book goes on to elaborate. We do not need to repeat them here. Rather, our interest is in this statement that time becomes another dimension of space. We know what every word in this statement means. It makes sense in ordinary English in some vague way that makes us think that we understand its content. But is any meaningful information being conveyed to the lay reader? The answer is probably 'no'. And the reason is interesting. Surely it is because here one encounters a scientific idea that as yet admits of no handy everyday analogy. One cannot say time becomes space is like a chrysalis becoming a butterfly or a water droplet evaporating or whatever. It is more like 'the sound of one hand clapping'. It is just a bare statement of the mathematical reality. This is the first occasion that this impasse has been encountered in my experience (others can probably come up with lesser known examples). It reveals why it is so difficult to get across the essence of this quantum cosmological picture of time at a popular level and why so many general readers found this a sticking point of the book. A reflection, perhaps, of Einstein's dictum that things should be made as simple as possible, but no simpler. However, the dilemma of the missing analogy may have a deeper significance.

One can take two views of the activities of scientists like physicists who are seeking things they call the fundamental laws of Nature. Either you believe, as they often do, that they are discovering the real thing and that one day we will hit upon the mathematical form of the ultimate laws of Nature. Alternatively, one may be more modest and regard the scientific enterprise as an editorial process in which we are constantly refining and updating our picture of reality using images and approximations that seem best fitted to the process. The latter perspective may of course ultimately converge upon the former one in the course of time. An interesting feature of the latter view is that it links the popularisation of science, and the analogical devices employed to propagate it, to the activity of scientific research itself. The constant updating of our scientific picture of how the Universe works can be viewed as a search for more and more sophisticated analogies for the true state of affairs. As they become more refined, they break down less and less often but, when sufficiently diluted, they become the homely images of the scientific populariser. Viewed in this light, readers of popular science come to see their activity as part of a continuous spectrum of imagery designed to make plain the workings of the world. They are closer to being actors than spectators.

The absence of a ready analogy for the central notions of new theories about the nature of time should not be viewed as a defect of those theories. On the contrary, it may be a healthy feature. Richard Feynman criticised some developments in fundamental physics for the enthusiasm with which they identified similar concepts and interrelationships existing in superficially quite different areas of physics. While some welcomed this analogy as evidence of the deep unity of Nature, Feynman preferred to see it as a sign of our impoverished imaginations. We just could not think up any new structures! It is just that 'when we see a new phenomenon we try to fit it into the framework we already have—until we have made enough experiments, we don't know that this doesn't work. So when some fool physicist gives a lecture at UCLA in 1983 and says: "This is the way it works, and look how wonderfully similar the theories are", it is not because Nature is *really* similar; it is because the physicists have only been able to think of the same damn thing, over and over again.' So when analogies from everyday experience are not readily available to illustrate new ideas in fundamental science it may be a good sign that we are touching some brute fact of reality that cares not one whit whether there exist analogues elsewhere in the world. When there are simple everyday analogies of what goes on in the first exotic quantum moments of the Big Bang the sceptic might be suspicious that we are importing our existing ideas rather than discovering new ones. Freeman Dyson once suggested that the trouble with the majority of hare-brained theories of elementary particles is not that they are crazy but that they are not crazy enough. True novelty and originality requires new images to be conceived, not merely the reprocessing of old ones. Perhaps the 'time becomes space' mantra is just radical enough.

Scientific explanation has come to be equated with mathematical explanation. Social scientists and other 'consumers' of mathematics are happy to regard it as a tool devised by humans that is useful. But for the fundamental physicist, mathematics is something that is altogether more persuasive. The further one goes from everyday experience and the local world, the correct apprehension of which is a prerequisite for our evolution and survival, the more impressively mathematics works. In the inner space of elementary particles or the outer space of astronomy the predictions of mathematics are almost unreasonably accurate. If one takes matter to pieces and probes to the root of what those pieces 'are' then ultimately we can say nothing more than that they are mathematics: they are relationships. Moreover it is not just the quantity of mathematics that imbues the structure of reality that is impressive, it is its quality. The sort of mathematics that lies at the heart of the general theory of relativity or of elementary particle physics is deep and difficult, far removed from the layperson's conception of mathematics as a form of high-level accountancy.

This has persuaded many physicists that the view that mathematics is simply a cultural creation is a woefully inadequate explanation of its existence and effectiveness in describing the world. If mathematics is discovered rather than invented—if 'pi' really is in the sky—then we can say something more about the analogical structure of the world. For, when we see the continuing refinement of our image of Nature in the development of more abstract mathematical theories of its workings, then, asymptotically, we are learning something about mathematics. If mathematics is just another human idiom that captures some but not all aspects of the world then it is yet another analogy that ultimately fails. But, if the world *is* mathematical at its deepest level then mathematics is the analogy that never breaks down.

A curiosity in this respect has been the investigation of the theory of superstrings. After early euphoria at the special nature of these 'theories of everything' the sober reality of the situation has emerged. Laws of Nature are much simpler that the outcomes of those laws that make manifest the world around us. This is a reflection of the fact that mathematical equations are much easier to find by appeal to general symmetry principles than they are to solve. While we may have the mathematical theory of superstrings, we cannot at present extract the predictions and explanations of the world that the theory has to offer. One of the foremost investigators of superstrings, Princeton physicist Ed Witten, has remarked that we may simply have been lucky enough to stumble upon this theory 50 years too early, before mathematics was sufficiently advanced to handle it. 'Off the shelf' mathematics has not been sufficient to unveil the secrets of superstrings. And, in fact, the quest for those secrets has pointed mathematics in new and fruitful directions of great novelty. Again, this lack of progress with a theory of physics may be a healthy sign of having taken the hard path that leads to the truth rather than the easy way of wishful thinking that seeks progress for its own sake. But while superstrings theory has its novel aspects—allowing additional dimensions of space and abandoning the traditional view that the most elementary constituents of matter are point-like—is it radical enough? Maybe, but if only superstrings weren't so much like little rubber bands.

CHAPTER 3

Falling between two cultures

You want to know the thought of Jacques Derrida.
There ain't no writer, there ain't no reader either.

Anonymous

The *New York Review of Books* once included an entry in its personal columns that read 'Petite, attractive, intelligent WSF, 30, fond of music, theatre, books, travel, seeks warm, affectionate, fun-loving man to share life's pleasures with view to lasting relationship. Send photograph. Please no biochemists'. If we were now asked to 'discuss' in that throwaway manner so beloved of those who set examination papers, what would we conclude? If we scour the book review pages of our national newspapers, how often do we ever find accounts of anything scientific? Yet the obscurest obscurantist's collected postcards from Eastbourne find a ready platform. And, if once in a while a work of popular science or biography breaks through the cultural divide, it will be treated as though it is necessary to transmogrify it ready for general consumption. It will be reviewed by a house reviewer with infinitesimal knowledge of the subject—Salman Rushdie on *A brief history of time* being a particularly curious case in point.

This cultural divide is a rather British tradition. The American media gives equal time to science and the arts. One is as likely to read about low-temperature physics as *Ice-station Zebra* on the books pages of the *New York Times*. This reveals a belief that the public both wants and needs to know about the development of science, both its ideas and applications. Not, of course, only science, to the exclusion of all those other things that our media think that we only want to know about, but in addition to them. Whereas, in this country, the book-review and literary pages of magazines and newspapers have become, by default, the non-science pages.

A review of *The Longman Literary Companion to Science*, edited by Walter Gratzer, Longman, October 1989

Many of the attempts made to bridge the literary and scientific gaps are a little clumsy—west-end plays about quantum mechanics, anthropic John Updike novels, science magazines illustrated with works of modern art, Zen and the art of quantum field theory, or the dissolution of the content of books under personal hype about their authors so that books get read about rather than read. It is easy to fall between two stools and end up with a product of the two that is neither good science nor good writing. Science fiction can totter uneasily here.

Walter Gratzer shows us that there is at least one better way. With a good knowledge of what has been written in the past and a lively imagination, it is possible to rise between the twin stools of literary interest and scientific content and display scientific writing that is good for its own sake. Gratzer has pulled together more than 200 pieces of writing, either by scientists or about science—some fact, some fiction, fictional science rather than science fiction, some poetry, some prose—to produce the sort of volume that would make long train journeys (and most of them are long) a more enjoyable experience. Indeed, what could be more agreeable than to imagine Einstein asking a porter at Paddington 'Does Oxford stop at this train?'

To be successful, books like this must be unpredictable and offbeat; and that, I suspect, requires a special type of editor: an anecdotalist rather than merely a generalist. The collection by Bernard Dixon is a perfect illustration of this theorem. It is a collection of pieces of good science writing, many by contemporary scientists. The pieces chosen are worth reading, although the avid devotee of the genre will surely have caught most of them first time around *in situ*. There are occasional pleasant surprises from the 19th century, some surprising omissions from the 20th—and how could Sir Arthur Eddington, to my mind the finest scientific writer of the 20th century, be left out of a book entitled 'classic writings in science'? But it was all so predictable. One set of railway lines is quite enough for a train journey. Are there no undiscovered by-ways, no peculiarities, no curiosities, no tales of the unexpected?

There are, and Gratzer has done a nice job in searching some of them out. It would be churlish to mention pieces that do not appear (this is meant to be a companion not an attractively bound millstone) but Stephen Leacock surely deserved a place with *Boarding House Geometry* ('The landlady of a boarding-house is a parallelogram—that is, an oblong figure, which cannot be described, but which is equal to anything') or C's journey to his final earthly resting place recounted in the saga of *A, B, and C* ('The funeral was plain and unostentatious. It differed in nothing from the ordinary, except that out of deference to sporting men and mathematicians, A engaged two hearses. Both vehicles started at the same time, B driving the one which bore the stable parallelopiped containing the last remains of his ill-fated friend. A, on the box of the empty hearse, generously consented to a handicap of a hundred yards, but arrived first at the cemetery by driving four times as fast as B. Find the distance to the cemetery.') Another little known missing masterpiece, and one that deserves to be read more widely in these days of constant hankering after technical progress and greater 'efficiency', is Arthur Clarke's 1951 short story *Superiority* about the defeat of an advanced civilization because of the inferior science of its enemies.

Of course, returning to where we began, it is hard to imagine works such as these drawing together highlights from non-scientific writing; they would be regarded as a sort of crammer for cocktail party small-talk about 'the arts' which save one the bother of reading the original books. But who cares: read these books and you might be able to amaze your friends with talk of the second law of thermodynamics—that's the one, you remember, that tells you that you can't win.

Part 2

Life in the Universe

If God had meant us to do philosophy, he would have created us.

Marek Kohn

Nothing is quite as eye catching as juxtaposing the words 'life' and 'Universe'. Inner and outer, familiar and strange, touchable and untouchable, the contrast could not be greater. It conjures up images of little green men, musings about the ultimate nature of existence, paranoid wondering about whether we are alone in the Universe, and strong feelings about what we should be doing about it: broadcasting our presence to all and sundry across the intergalactic airwaves (no air there, but you know what I mean) or keeping our heads down. So far, we have found no strong evidence of life beyond Earth, but this could be a consequence of many quite different situations within our galaxy. We might indeed be alone, the only technologically advanced species within signalling range. This could be because we got there first or because advanced extraterrestrials are very rare. Soon after they develop the know-how to create weapons of mass destruction, they may kill themselves off by global warfare, accidents with nuclear technology, pollution, disease, be annihilated by asteroidal impacts, or simply run out of space and raw materials. Alternatively, life may be very common and we are just a rather uninteresting, run of the mill example of which there are millions in the Milky Way galaxy. The most advanced civilisations are not terribly interested in another example of something they have seen many times before. Then again, we may be rather interesting, but, just as we like to leave rare animals alone on game reserves, we may have been left to evolve without interference. Or, perhaps those extraterrestrials that talk to one another do so using a form of technology that is way beyond our present scientific understanding. In this way they ensure that a very high level of intelligence and maturity is a prerequisite for joining the galactic club. Only those civilisations that have overcome the great problems that we face have graduated to take part in those intergalactic dialogues.

At first sight it appears that nothing could be less concerned with life on

Earth, or anywhere else for that matter, than this thing we call the Universe. The Universe is big and old, dark and cold: life is small, recent, fragile, and fleeting, requiring warmth and light for its germination and sustenance. What is the connection? For hundreds of years humanistic thinkers have denied that the Universe could be anything but hostile to the presence of life, but modern cosmology has put a strange spin on these statements. The discovery, only made in the late 1920s, that the Universe is expanding, means that its size is bound up with its age. We know that the Universe is more than 12 billion years old, but its great age is no coincidence. It takes billions of years to form the basic building blocks needed for any form of chemical complexity like the phenomenon we call 'life'. These blocks are built up by a slow sequence of nuclear reactions inside the stars: from hydrogen to helium, from helium to beryllium, to carbon, oxygen, and beyond. This 20th-century insight teaches us that the Universe must be billions of years old if it is to be possible for living observers to evolve, and so it must also be billions of light years in size. And, by the same token, the vast amount of expansion that is required must dilute the density of matter and radiation in the Universe to such an extent that stars and galaxies are vastly separated from one another and the night sky is dark. If life evolves it will most likely be isolated from other inhabitants of the Universe, watching the dark night sky, and impressed by the ineffable vastness of space.

This section contains articles that discuss some of the unexpected connections between the conditions needed for life to exist on a planet like Earth and the structure of the Universe as a whole, the form of its laws, and the unchanging 'constants' of Nature that define the fabric of the world around us. The background to this is that astronomers have discovered that our existence was a close-run thing. If the Universe had been only very slightly different then living beings like ourselves could not have existed. Perhaps no living beings of any sort could have existed because there are small changes that could occur that would prevent any atoms existing at all. Alternatively, if there are small variations in the quantities that we call the 'constants of Nature' from one place to another in the Universe then some places will be more hospitable to the evolution of life than others. Modern theories of quantum physics and gravity lead us to expect that there will inevitably be small variations of this sort imprinted upon the Universe when its expansion was beginning.

This 'anthropic' perspective upon the structure of the Universe is the subject of the first chapter in this section '*Anthropic principles in cosmology*', and there is more in the short piece '*Cosmology*', based on a talk about modern cosmology given at the Venice Film Festival, to provide a modern astronomical context for Olmi's film *Genesis*. '*Long-distance calls*' highlights what this perspective has to teach us about the existence and likely inclinations of any other sentient beings that might be out there looking at the Universe.

As we look more closely at the nature of living things on Earth we can isolate a number of fascinating features of their design and behaviour that are direct consequences of their size in all its manifestations: length, height, surface, volume, and weight. In '*Sizing up the Universe*', we see how the sizes of the things around us on Earth, and in the heavens, are by no means as haphazard as they might at first appear. Habitable planets must be big enough for their gravity fields to hang on to an atmosphere, but not so big that their gravity is strong enough to break the bonds between molecules. Of course, physics and astronomy can only define the constraints within which biological diversity and evolution has to work. Two of the biggest current questions about life concern the extent of the influence of natural selection. Does it fashion things like our aesthetic preferences or emotions to any significant extent, or is its role confined to shaping the form of our physical bodies? Do the laws that govern the development of complexity play a significant role in the appearance of life, or can it be explained by nothing more than evolution by natural selection? In '*The truth is in the choosing*' we take a look at attempts to use evolution as a tool to learn about why our thinking about the Universe is reliable. Many scientists have begun to use the insights of evolutionary biology to attempt to understand why our minds have many of the inclinations that they do. If possession of a certain propensity is more likely to lead to survival or fecundity then it will tend to be preferentially passed on to our children. However, the difficulty is that the passing on was mainly done millions of years ago, in environments quite dissimilar to those in which we now live. Our own behavioural biases are therefore likely to be by-products of adaptations to environmental conditions that no longer necessarily exist.

CHAPTER 4

Anthropic principles in cosmology

The Universe; a device contrived for the perpetual astonishment of astronomers.

Arthur C. Clarke

Where is there 'life'?

Life is a manifestation of the attainment of a particular level of organised complexity in a physical system. This is not all that it is, but it identifies some of the necessary conditions for its evolution and persistence. Our knowledge of living things is limited to those to be found within the terrestrial biosphere. Despite the powerful, and not always positive, influence of these terrestrial organisms upon the character and evolution of the surface and atmosphere of the planet, they are little more than superficial additions to the make-up of the Earth. The quantitative situation is summarised as follows, with the masses of different varieties of living material given, for convenience, in units of 1 Pg = 10^{15} gm, which is about the mass of a large mountain. The total mass of dry living material (organisms like animals, land-based plants and trees) on the Earth's surface amounts to 1841 Pg, only about 10^{-9} of the mass of the Earth. The marine biosphere contributes a mass of 4 Pg. Of the total, 1791 Pg (97.3%) is plant life whilst 50 Pg (2.7%) is animal life of which 30 Pg is in the form of insects. There are more than 350,000 plant species and more than 1,200,000 animal species, of which over 800,000 are insects; in addition there are over 100,000 species of micro-organism.[1]

If we look further afield we can place the terrestrial biosphere in a cosmic context in terms of its size and mass. Figure 4.1 displays all the most significant structures to be found in the Universe, from the subatomic world to great clusters of galaxies and the extent of the visible part of the Universe—that is, the distance light has travelled since the expansion of the Universe began about 15×10^9 years ago. The region of the size–mass diagram within which living things are found is indicated. We see that it is limited to relatively small dimen-

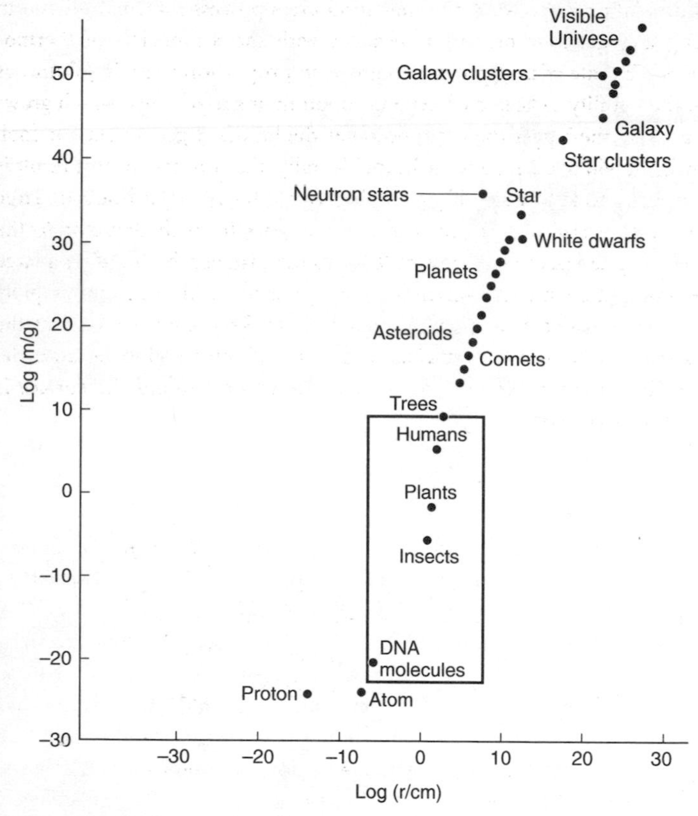

Fig. 4.1. The distribution of notable cosmic objects by size and mass. The box indicates the region in which one finds examples of organised complexity.[2]

sions and masses. Examples of organised complexity found outside this box are dynamical in nature; that is the complexity exists in their motions, not in their structure. To a considerable extent this is not surprising: size is an important determining factor in the evolution and persistence of living things. Galileo was the first to exhibit a clear appreciation of the fact that as structures grow in volume, in proportion to R^3, if R is a characteristic dimension, their strength grows only as R^2 because they fracture across cross-sectional areas. Hence, if one just scales up living things in size they become unsupportable and break by compressional or bending moment stresses when these exceed the strength of intermolecular bonds. Get too big and you break! This is why trees and other living things are limited in size; humans are the largest bipedal living creatures.[3] Similarly, if one wants to build a 'brain', whether it is living or artificial, then size again plays a key role. Brains and computers process information and in so doing they generate heat in accordance with the second law of thermodynamics. The rate of heat generation grows in proportion to their volume, as R^3, but their ability to keep cool depends upon their surface area, which grows only as R^2. So the bigger the brain becomes the harder it gets to keep it cool. Eventually, it will overheat and melt. Incidentally, the converse of this result is that it is easier to keep warm if you are big. Accordingly, polar bears are large and the average size of birds increases as one goes from the equator to the poles. If one attempts to circumvent this cooling problem by building a large brain or computer with a fractal surface area that grows with scale more rapidly than R^2 (like a towel or a sponge), then one can keep cool but only at the expense of becoming a less efficient processor of information because the length of the circuitry required to connect the points around the surface is enormously increased.[4]

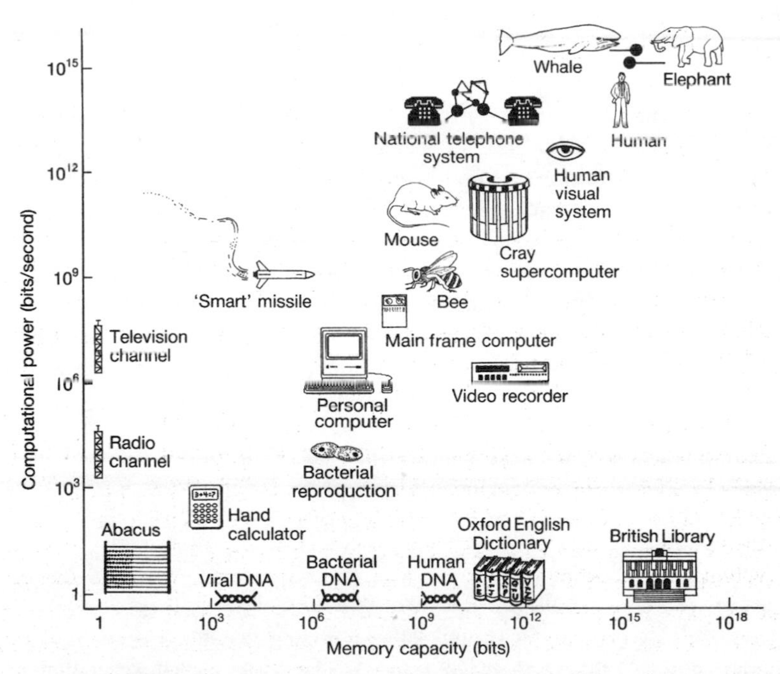

Fig. 4.2. A display,[2] adapted from ref. 5, showing complex organised structures in terms of their volume (calibrated by their capacity for storing information) and processing power (measured by the speed at which they can change that information). We recognise the most complex structures as those at the top right of the graph.

What is life?

The structures that lie within the box in Figure 4.1 are what they are, not because of their size, or what they are made of, but because of the ways in which their components are *organised*. This aspect is illustrated in Figure 4.2 where organised complexity is characterised by size, in terms of information storage capacity, and processing power. Some structures, like books or library archives, store very large quantities of information but do not have the ability to transform it into new information. Computers, living things and many other systems shown in the figure do have that ability. The most sophisticated complex structures are those towards the top, right-hand corner of the graph, which combine the ability to store large quantities of information with the ability to process it rapidly. There is an infinity of possible structures stretching up this diagonal, but whether or not there exist new mathematical principles governing the existence and stability of complexity in general is still an open question.

Before leaving these illustrations of living things and their place in the universe of possibilities one should reiterate that the most interesting structures in Figure 4.2 are not what they are, because of their constituents alone. Moreover, they exhibit a number of characteristic features, including self-organisation, non-linear feedback and local teleonomy. They exhibit this structural complexity because of the ways in which their constituents are organised. This organised complexity makes them more than the sum of their parts. It also means that the reductionist philosophy is severely limited. We cannot understand a computer or a bird just by knowing what it is made of; we need to know how its components are hard-wired together. Knowledge of the ultimate *laws* of Nature is of no great help in this respect because complex structures are *outcomes* of the laws of Nature that do not possess the same symmetries as the laws themselves. The underlying symmetries are broken in the outcomes. This is why a world governed by simple laws can be so complicated and why a knowledge of the laws of physics is necessary but not sufficient to understand the structure of the Universe around us.[2]

Universes that can contain complexity

Let us turn now from the physical characterisation of living things as complex organised structures, to the cosmological conditions that are necessary for the existence of such structures. Biologists believe that the spontaneous evolution of organised complexity requires the presence of carbon atoms. This is not to say that complexity cannot have other atomic foundations, only that these require catalysis by the presence of other forms of carbon-based complexity in order for their appearance to be probable.[3] For example, we might highlight ourselves as a particular form of carbon-based complexity that is able to bring into existence other forms of silicon-based 'artificial' intelligence by our intervention.

Modern astrophysics has taught us that all the elements heavier than helium have their origins in nuclear reactions within stellar interiors. These nuclear fusion reactions build the biologically interesting elements, like carbon, oxygen, nitrogen, phosphorus (and even silicon), from nuclei of hydrogen and helium produced during the first three minutes of the Big Bang. But this stellar production of carbon takes billions of years to complete. When stars have completed their main sequence lifetime they undergo a period of explosive evolution that produces and disperses the biological elements, such as carbon, around the Universe. Eventually, that carbon condenses into solid material and finds its way into planets and, ultimately, into people. Thus we can begin to appreciate why a Universe that contains life needs to be so large. In order for it to contain the basic atomic building blocks of biological complexity, it must be old enough for the stars to have produced them. This takes billions of years and so, since the Universe is expanding, it must be billions of light years in size. A Universe that was only as big as the Milky Way galaxy would be little more than a month old even though it might contain a hundred billion solar systems.

This teaches us that we should only expect to find living observers in a

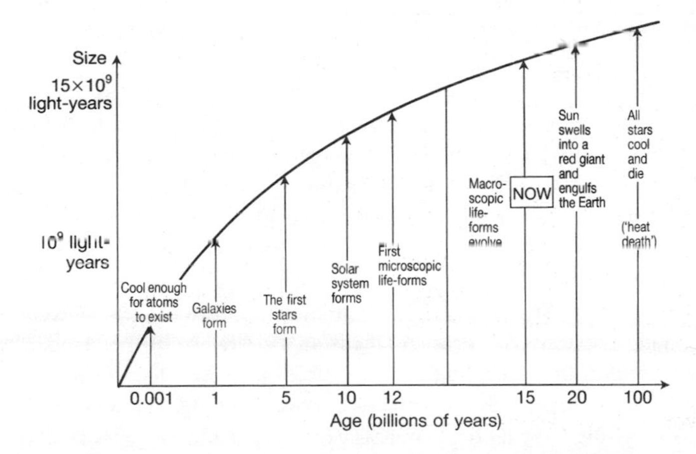

Fig. 4.3. The overall pattern of cosmic theory from the time when the Universe was about a million years old until the far future. The expansion ensures that the ambient temperature of the Universe is steadily falling. Only after a particular period of time are conditions cool enough for the formation of atoms, molecules, and living complexity. To the future we expect an epoch when all stars will have cooled and died. Thus, there exists a niche of cosmic history before which life cannot evolve and after which conditions do not exist for its spontaneous evolution and persistence.

Universe after a period of several billion years, and after the stars have formed. We might also argue that if life has not evolved before all the stars have exhausted their nuclear fuel and died then the probability of its subsequent evolution is very small. So there is an interval of cosmic history before which life, and after which living complexity, cannot evolve. It is therefore not surprising that we find ourselves living within this niche of cosmic history as shown in Figure 4.3.

The important lesson that we learn from these considerations is that our view of the Universe is biased by our own existence. We see the Universe when it is relatively old and cold because living observers cannot exist at other times. Clearly, such a temporal bias upon our cosmological perspective can be extended to our position in space. If our (possibly infinite) Universe is very different in structure from place to place then we must find ourselves living in one of those regions where conditions satisfy the necessary conditions for the existence and persistence of life.

The inflationary universe

This perspective upon the link between the structure of the Universe, and those aspects of its structure that are necessary conditions for the emergence of complexity within it, plays a very important part in our evaluation of the correctness of cosmological theories in which there is any random element in the structure and development of the very early Universe. When that is the case the local properties of the Universe (features like its density, temperature, matter–antimatter balance) can vary from place to place, and it is even possible for the constants of Nature or the number of dimensions of space to be random variables, according to current theories of high-energy physics. The fact that we necessarily inhabit a region that satisfies conditions on the local properties and values of the local constants that permit the existence of life then becomes an important factor in weighing the likelihood of any theory being true, given the observational facts.

One example of this state of affairs is provided by the most general edition of the inflationary universe theory.[6] This is an elaboration of the standard Big Bang picture of the expanding Universe which proposes that during the very early stages the expansion *accelerated* for a brief period.[7] The standard (non-inflationary) Big Bang model manifests expansion that decelerates at all times, regardless of whether the expansion continues forever or collapses to a big crunch in the future. This brief period of inflation has a variety of interesting consequences: for example, it can guarantee that the visible Universe possesses many of the mysterious properties that are observed. This accelerated phase will arise if certain types of matter exist during the very hot early stages of the Universe. If this matter (which we call the 'inflaton field') moves slowly enough

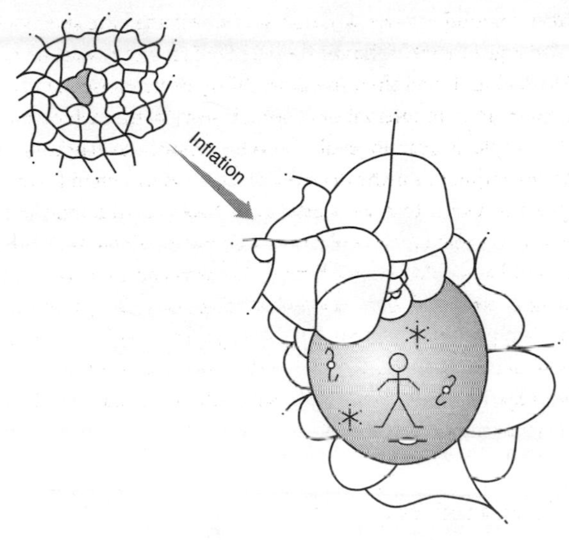

Fig. 4.4. Small, causally connected domains existing near the beginning of the expanding universe inflate separately by different amounts. We must find ourselves in one of those domains that expands to more than about ten billion light years in size. Only these large, old domains have enough space and time for the evolution of stars and the ensuing production of the elements heavier than helium, which are necessary for the evolution of complex structures.

then it drives a period of inflationary expansion. At the very early time when this is expected to occur light will have had time to travel only about 10^{-25} cm since the expansion began. Thus, within each spherical region of this diameter there should exist a smooth correlated environment, but conditions could be quite different over more widely separated dimensions because there will not have been enough time for signals to travel between different regions. This situation is represented in Figure 4.4.

In each of these causally disjoint regions the inflaton field will evolve differently, reflecting the differing conditions and starting states within the different regions. As a result, different regions will accelerate, or 'inflate', by different amounts in some random fashion.[8] Some regions will inflate by large amounts and give rise to regions that expand for more than ten billion years. Only in these regions will the formation and evolution of stars, of heavy elements and of organised complexity be possible.[3] We necessarily live in one of those regions. When we come to evaluate the correctness of any particular theory of high-energy physics leading to inflation we have to take into account that even if the unconditional probability of producing a large domain is small we would have to find ourselves within it.

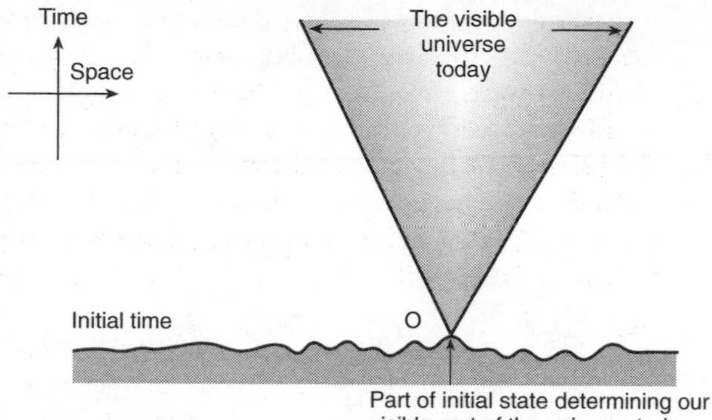

Fig. 4.5. A space and time diagram showing the development of our visible universe from its initial conditions. The structure of our entire visible universe is determined by conditions at the point 'O' where our world-line intersects the spatial surface of zero time where the density is assumed to be infinite. The conditions at O may need to be atypical to permit the subsequent evolution of observers, and their detailed local form will not be determined by any global principle that determines the initial conditions statistically.

The visible universe

We should draw a distinction between '*the Universe*', which might be infinite in extent, and the '*visible universe*'—that part of the Universe from which we have had time to receive light signals since the beginning of the expansion, about fifteen billion years ago. The visible universe is finite and steadily getting bigger as the Universe ages. All our observations of the Universe are confined to that subset of the whole that is the visible part. Now, take the region constituting the visible universe today, reverse the expansion of the Universe and determine the size of that region which, at any given time in the past, is going to expand to become our visible Universe.[1] When the Universe is about 10^{-35} s old the present visible universe was contained within a sphere less than one centimetre in radius. This state of affairs has a number of important consequences for any quest for a 'Theory of Everything'. In particular, as can be seen from Figure 5, our visible universe is the expanded image of an infinitesimal part of the initial conditions as a whole.

We can never know what the entire initial structure of space looks like. Moreover, while it is fashionable amongst cosmologists to attempt to formulate grand principles that specify the initial conditions; for example, Roger Penrose's minimum Weyl entropy condition,[9] Alex Vilenkin's 'outgoing wave condition'[10], or Stephen Hawking and Jim Hartle's 'no boundary condition',[11,12] these principles specify mean conditions over the entire initial data space. Ultimately, these conditions will be quantum statistical in character. However, such a principle could never be tested even with all the information available in the visible universe. Moreover, the entire visible universe is determined by a tiny part of the initial data space, which may well be atypical in certain respects in order that it satisfy the conditions necessary for the subsequent evolution of observers. Global principles about the initial state of the Universe, even if correct, may be of little use in understanding the structure of the visible part of the Universe because the Universe may evolve from an idiosyncratic part of an infinite span of initial conditions.

Constants of Nature

The constants of physics, quantities like the mass of an electron or its electric charge, have traditionally been regarded as unchanging attributes of the Universe. An acid test of any 'Theory of Everything' would be its ability to predict the values of these constants.[13]

The values of the constants of Nature are what endow the Universe with its coarse-grained structure. The sizes of stars and planets, and to some extent of people also, are determined by the values of these constants.[2,14] In addition, it has gradually been appreciated that the necessary conditions for the evolution of organised complexity in the Universe are dependent upon a large number of remarkable 'coincidences' between the values of different constants. Were the constants governing electromagnetic or nuclear interactions to be very slightly altered in value then the chain of nuclear reactions that produce carbon in the Universe would fail to do so; change them a little more and neither atoms nor nuclei nor stable stars could exist.[2] So, we believe that only those universes in which the constants of Nature lie in a very narrow range (including, of course, the actual values) can give rise to observers of any sort.

In recent years attempts to develop a quantum cosmological model have uncovered the remarkable possibility that the constants of Nature may be predictable but their values will be affected quantum statistically by the space time fabric of the Universe.[15–17]

If we relinquish the idea that the topological structure of the Universe is a smooth ball then we picture it as a crenellated structure with a network of wormholes connecting it to itself and to other extended regions of space–time, as pictured in Figure 4.6. The size of the throats of the wormholes is of the order of 10^{-33} cm. Now, this turns out to be something more interesting than simply generalisation for its own sake. It appears that the values of the observed constants of Nature on each of the large regions of space–time may be determined quantum statistically by the network of wormhole connections. Thus, even if those constants had their values determined initially by the logical strait-jacket of some unique 'Theory of Everything' (like superstrings[18]), those values would be shifted so that today they would be observed to have values

Fig. 4.6. Wormholes connecting different large subregions of the Universe to themselves by wormhole 'handles' and to other such regions by wormhole 'throats'.

given by a calculable quantum probability distribution. However, there is a subtlety here. We might think that it is the business of science to compare the most probable value predicted with that observed. But why should the observed universe display the most probable value in any sense of the word 'probable'? We know that observers can only evolve if the constants lie in a very narrow range, so the situation will be as shown in Figure 4.7.

We are interested, not in the most probable value of a constant, but only in the conditional probability of the constant taking a value that subsequently per-

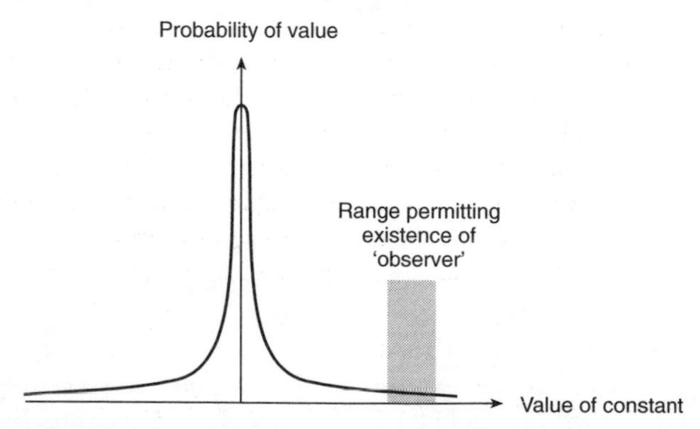

Fig. 4.7. A hypothetical prediction for the measured probability of a constant of Nature when the Universe is large and old (i.e. now), but the small range of values for which living complexity can exist is also shown. The latter range is far from the most probable value predicted by the wormhole fluctuations. However, we are only interested in the most probable value given that observers can subsequently evolve.

mits living observers to evolve. This may be very different to the unconditioned 'most probably' value.

The number of dimensions of space

This type of argument can be extended to consider the dimensionality of the part of space that becomes large.

Superstring theories naturally predict that the Universe possesses more than three dimensions of space,[18] but it is expected that only some of these dimensions will expand to become very large. The rest must remain confined as imperceptibly small. Now we might ask why it should be *three* dimensions that grow large. Is this the only logical possibility or just a random symmetry breaking that could have fallen out in another way, or perhaps *has* fallen out in different ways all over the Universe so that regions beyond our horizon may have four or more large dimensions? Again, we know that the dimensionality of space is critically linked to the likelihood of observers evolving within it,[19] (for instance, there are no atoms or stable gravitationally bound orbits in more than three dimensions) and similar considerations to those we have just introduced with regard to the probabilistic prediction of the constants apply to any random symmetry-breaking process that determines the number of space dimensions that become large.

Conclusions

These examples teach us that if there exists any random element in the initial structure of the Universe, or its early evolution (and quantum theory makes this random element inevitable), then we observe aspects of things that are not typical. They are conditioned by the necessary conditions for the evolution of biochemical complexity. Thus, remarkably, we discover that a full appreciation of those aspects of the Universe and its laws that play a role in the possibility of living observers is necessary if we are to interpret cosmological observations correctly and draw from them the correct conclusions regarding the nature of initial cosmological conditions and the ultimate form of the laws of Nature.

References

[1] M. Taube, *Evolution of matter and energy on a cosmic and planetary scale*, Springer (1985).

[2] J.D. Barrow, *Theories of everything: the quest for ultimate explanation*, Oxford University Press (1991).

[3] J.D. Barrow and F.J. Tipler, *The anthropic cosmological principle*, Oxford University Press (1986).

[4] J.D. Barrow, *The world within the world*, Oxford University Press (1988).
[5] H. Moravec, *Mind children*, Harvard University Press (1988).
[6] A. Guth, 'The inflationary universe: a possible solution to the horizon and flatness problems', *Phys. Rev. D* **23**, 347 (1981).
[7] J.D. Barrow, 'The inflationary universe: modern developments', *Q. J. R. Astron. Soc.* **29**, 101 (1988).
[8] A. Linde, *Particle physics and inflationary cosmology*, Gordon and Breach (1989).
[9] R. Penrose, *The emperor's new mind*, Oxford University Press (1989).
[10] A. Vilenkin, 'Boundary conditions in quantum cosmology', *Phys. Rev. D* **33**, 3560 (1982).
[11] J.B. Hartle and S.W. Hawking, 'The wave function of the universe', *Phys. Rev. D* **28**, 2960 (1983).
[12] S.W. Hawking, *A brief history of time*, Bantam (1988).
[13] J.D. Barrow, 'The mysterious law of large numbers', in *Modern cosmology in retrospect*, eds S. Bergia and B. Bertotti, Cambridge University Press (1990)
[14] B.J. Carr and M.J. Rees, 'The anthropic principle and the structure of the physical world', *Nature* **278**, 605 (1979).
[15] S. Coleman, 'Why there is something rather than nothing', *Nucl. Phys. B* **310**, 743 (1988).
[16] S.W. Hawking, 'Wormholes in space–time', *Phys. Rev. D* **37**, 904 (1988).
[17] S.W. Hawking, 'Baby universes', *Mod. Phys. Letts* **A 5**, 453 (1990).
[18] M. Green, J. Schwarz and E. Witten, *Superstring Theory*, (2 vols), Cambridge University Press (1987).
[19] J.D. Barrow, 'Dimensionality', *Phil. Trans. R. Soc. A* **310**, 337 (1983).

CHAPTER 5

Sizing up the Universe

Give me the ninety-two elements and I'll give you a universe. Ubiquitous hydrogen. Standoffish helium, Spooky boron. No-nonsense carbon. Promiscuous oxygen. Faithful iron. Mysterious phosphorous. Exotic xenon. Brash tin. Slippery mercury. Heavy-footed lead.

Chet Raymo

The world around us is filled with a multitude of things of all shapes and sizes. We sit midway between the vastness of cosmic dimensions spanning interstellar space and the microcosm within the atoms that comprise us.

The range of sizes

Have you ever wondered why things are the size they are? Could trees be a hundred times bigger? How small can a star be? Is there any limit to the size of a computer? And how does the size of things influence our daily lives?

Let's begin by doing a bit of armchair fieldwork. Draw a graph of size against mass and mark on it the positions of as many different types of object as you can (Figure 5.1). Begin with very small subatomic particles like the proton and go all the way up to the largest galaxies and clusters of galaxies. Call a halt when you reach the scale of the whole 'visible' universe. In between the inner space of the subatomic particles and the expanse of the whole visible universe we find an immense range of things: first, there are clusters of galaxies, individual galaxies, star clusters, stars, planets and moons, asteroids and comets, then, living things like trees and animals, fish, birds and insects, before we reach the microworld of bacteria, molecules and single atoms. When we see them all together on the size–mass figure the result is curious. They are not spread all over the figure in haphazard fashion. Large regions of the picture are completely empty and most things lie in a band running diagonally from the bottom left to the top right.

The structures that we see in the Universe result from a balance between the opposing forces of Nature. In clusters of galaxies and spiral galaxies, that balance is between the force of gravity pulling stars in towards the centre and

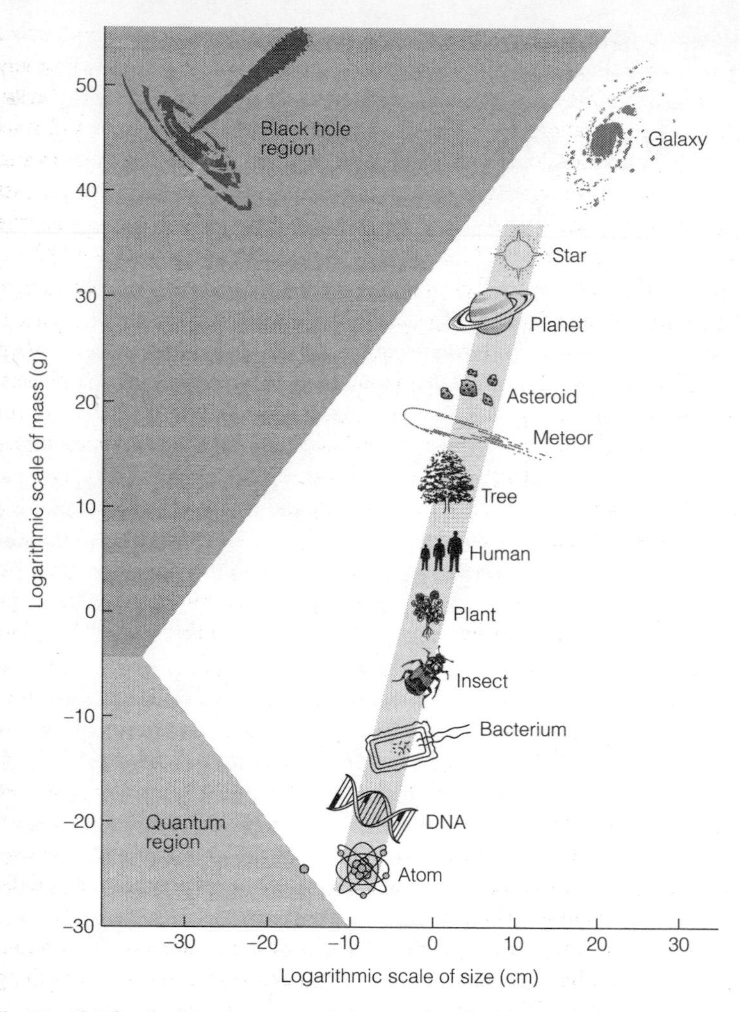

Fig. 5.1. The distribution of masses and sizes supplemented by a line of constant density equal to that of solid atomic structures, a line bounding the region populated by black holes, and a line bounding the region in which quantum uncertainty renders objects unobservable.

the outward reaction caused by the rotation of the stars around the centre. In individual stars the balance is between the inward squeezing of gravity and the outward pressure of gas or radiation produced by the nuclear reactions at their centres. Bodies that are too small for the gravitational squeezing at their centres to attain the temperatures of millions of degrees that are needed to initiate nuclear reactions will never become stars. They will remain the cold bodies that we call 'planets'. Planets are balancing acts in which atomic forces resist any tendency for atoms to overlap and thus counter the inward crushing by gravity.

These considerations have revealed to astronomers why planets and stars range in size as they do. However, they still do not know whether galaxies and galaxy clusters owe their sizes to a similar principle or whether they are just residues of haphazard irregularities that came into being along with the Universe. Whereas we would have predicted that stars and planets would exist, even if we had never seen either, we would not have predicted that the Universe would contain things like spiral galaxies.

A balance between the forces of gravity and atomic forces occurs when matter has a density close to the density of single atoms. This is because planets, mountains, trees and people are all composed of closely packed arrays of atoms. Their similar density means that their mass divided by their volume is about the same. Since volume is proportional to size cubed we see why these solid objects all lie along a line with a slope close to three in the figure. This is a line of constant density—atomic density. It extends all the way from the simplest atom of hydrogen (which consists of one proton and one electron) up to the largest solid structures. Star clusters, galaxies and clusters of galaxies are collections of orbiting stars, rather than solid objects, so they lie slightly below the atomic density line.

Black holes and quantum uncertainty

What about the empty spaces on our figure? The region of small size and large mass is completely empty. It is the realm of 'black holes'. These are regions of space that contain so much mass that the pull of gravity prevents anything, even light, from escaping. Anything in that part of our graph would be invisible, trapped inside a black hole.

If we probe down into the subatomic realm we encounter another limit on what we can see, however sensitive our instruments. When we 'see' something we record a photon of light that bounces off it into our eye or other detector. If the thing we are looking at is large then the recoil it feels from that photon is totally negligible. But for very small objects the effect of the recoil can produce a large disruption of what we are trying to measure. This catch-22 is expressed by the famous 'Uncertainty Principle', first elucidated by Werner Hesienberg: we cannot simultaneously measure the position and the speed of something to ever-increasing accuracy, no matter how perfect our instruments. Heisenberg's limit is shown on the figure. It reveals why another portion is empty (the 'quantum region'). Nature is constructed so that we cannot 'see' any objects in the bottom triangle of the picture without disrupting them by the very act of observation.

Our first important lesson is that the sizes of things in the Universe are not random. They lie in the ranges they do because they are manifestations of balances between the different opposing forces of Nature. Moreover, the existence of black holes and the 'Uncertainty Principle' dramatically limits the range of masses and sizes that can be seen.

Life on Earth

Let's move a little closer to home. We have just seen that there are some regions of the Universe where so much mass lies in so small a volume that the resulting pull of gravity prevents everything, even light, from escaping. We are familiar with a more modest example of this gravitational pull. If we throw a stone in the air then the Earth's gravity brings it back to the ground. Hurl things faster and they travel further before returning. But if we launch a rocket fast enough —faster than about 11 kilometres per second—then it can escape the Earth's pull completely. The bigger the planet the more material it contains, the greater the pull of its gravity, and the faster you need to go to escape from it into outer space.

The Earth and the Moon are totally different. The Earth is alive, with a biosphere of great complexity. The Moon is arid and dead. The reason for this is that the Earth has an atmosphere of gases like nitrogen and oxygen that promote intricate chemical and biological processes on its surface; the Moon has no atmosphere. The Moon is too small for its gravity to hang on to an atmosphere like the Earth's. If the air in a room were transported to the Moon the molecules would find themselves moving fast enough to escape into space. So, not only are the sizes of planets fixed by the balances between Nature's forces but those that could possess atmospheres and be habitable must be bigger than a critical size. Yet they must not be too big. Why?

Living things are made out of intricately organised collections of atoms and molecules that are held together by interatomic and molecular bonds. These bonds can be broken and when they are, matter changes, often irreversibly. Fry an egg and you see that the protein molecules in the egg-white are very mobile and flexible when the temperature is low, but as the temperature rises they become rigid and inflexible as the egg-white becomes solid and the egg is cooked. 'Cooking' means attaining the temperature at which intermolecular bonds are changed or 'denatured' as chemists say. All living things are made of atoms and molecules that are bonded in this way. Place them on a planet that is too big and the strength of gravity on its surface will be able to break those bonds. Life made from atoms could not exist on a planet that was too big.

This need for atomic and molecular bonds to hold together teaches us why mountains cannot have heights greatly in excess of Everest. As a mountain gets higher and heavier so the pressure on its base increases. If it becomes too large then the bonds between atoms start to break and the mountain just sinks into the planet surface.

There is a distinctive region of our size–mass figure where complicated living things reside. Living things, from trees to birds, land animals to sea creatures, are also strongly constrained in size. A tree that grows too large would suffer unacceptable pressures on its base and break. In practice, it is the susceptibility to breaking when slightly bent in the wind that snaps the tree and limits its maximum height. The problem is that the tree's strength does not keep pace with its size as it grows. This can also be seen in a more homely context. A kitten's tail will stand bolt upright like a spike. The small kitten is strong enough to support its little tail. But look at its bigger mother. Her longer tail loops over because she is not strong enough to hold it upright.

Strength and size: the problem of being big

This failure of strength to keep pace with increasing size and volume is evident if we watch creatures of different sizes. An ant can carry a load that is ten times greater than its own bodyweight. A small dog can carry another dog on its back without any trouble. A child can carry another child piggyback style without too much strain, an adult has much more difficulty and no horse is strong enough to carry another horse on its back. As you get bigger, the stress on your bones gets greater; they need to be larger and broader to stand the strain. The largest dinosaur was the Brontosaurus and at 85 tons it was pretty close to the size limit for a land-going animal. For comparison, the largest animal today is the African elephant, at about 7 tons. Dinosaurs managed to spread the load out over their widely spaced legs but if they fell their bones would almost certainly break. Adult humans have a much shorter distance to fall if they stumble but sometimes they still break bones. Children have an even shorter distance to fall and so are better protected against breakage. The lesson seems to be that if you want to be big you need some extra support. The place to look for that support is in the water. It is no accident that the largest blue whales (at 130 tons) are much much bigger than the largest land-going dinosaurs ever were. When any object is placed in water it feels a buoyancy force pushing upwards. The downward pressure of gravity is reduced and this is one of the reasons why pregnant women are encouraged to include regular swimming in their antenatal programme.

Water can also support you if you are small. Place a paper clip on the undisturbed surface of some water and it is supported by surface tension. As you add detergent to the water this force is reduced and the clip will sink. Very small creatures, like pond-skaters, can use surface tension to support their weight, as long as their legs spread out over a few millimetres. We would need to have our legs spread out over about 7 kilometres to achieve the same result!

If you want to fly then it also pays to be small. You need to generate enough flapping power with your wings to overcome the pull of gravity. As you get bigger the power required grows faster than does the muscle power that you can exert. The largest birds that can hover for long periods in still air are humming birds which vary weight from about 2 to 20 grams. In fact, they can even take off vertically. Of course, there are bigger birds. The biggest are Kori bustards, weighing in at about 12 kg; but they stay aloft by soaring and exploiting wind gradients or thermal currents. When you see a kestrel apparently hovering over a fixed point on the ground it is not hovering, it is flying against the wind so that its speed relative to the ground is zero. It is not strong enough to support its own weight by hovering in still air.

Size and culture

Clearly, our size has influenced many aspects of our social development in ways that are both good and bad. We are large enough and strong enough to be able to wield tools that can split rocks. But we can also use those tools as weapons that transfer enough momentum to kill another person. Our ability to use fire is also connected to our size. There is a smallest possible flame that can burn in air, because the surface area surrounding a volume of burning material determines the influx of oxygen required to sustain the combustion. As the volume decreases the surface area falls faster and the fire is eventually starved of oxygen. This is why insects cannot make use of fire in controlled circumstances. An ant just could not get close enough to refuel the fire without getting burnt.

One of the most important developments in human history was that of the written word and the use of paper and other lightweight materials for its representation. To make use of materials like paper it is necessary to be large. Small creatures, like flies, exploit the surface forces between molecules because they are stronger than gravity. For larger creatures gravity wins, and you cannot walk upside down on the ceiling. But those surface forces that help tiny creatures defy gravity ensure that they cannot manipulate surfaces. They could not turn the pages of microscopic books.

Surface area

Watch a rolling snowball as it grows bigger. As its radius increases so its volume and surface area grow as well. But whereas its volume grows as the cube of its radius, its surface area grows only as the square of its radius. So its surface area does not keep pace with its growth in volume. This losing battle of surface area against volume is another vital constraint upon the size of living things. As your volume increases with growth so your heart and other heat-generating organs increase in volume and output. But the ability to keep cool depends upon how much surface area you have from which heat can escape. Small creatures have a relatively large surface area for their volume; large ones have a relatively small surface area for their volume. In cold climates small creatures will be unable to generate enough heat from eating in order to keep warm; large animals will. So we find large animals, like polar bears, at the Poles rather than small mice, and the average size of birds increases as one goes from the equator to the poles. Small animals tend to huddle together for warmth. This is a way of reducing their heat loss because less of their surface is exposed.

The exposure of surface area also determines how fast materials will burn because the surface is where oxygen feeds the fire. Burning only occurs at the surface. Small objects expose relatively more surface and this is why wood-shavings 1 cm in size will burn many times faster than a log 10 cm in diameter.

Fractals

The study of surfaces that enclose a particular volume of space by a huge surface area is very fashionable amongst mathematicians. Such surfaces are examples of 'fractals'. They can be constructed by copying a basic pattern over and over again on a smaller and smaller scale. Spectacular fractal images can be found on posters and magazine covers—there have even been exhibitions of this computer 'art' at major international galleries—but there are more serious applications as well. The intricacy of fractal designs creates a way to increase the capacity of computer memory.

We see that Nature uses fractals everywhere: in the branching of trees and the shaping of leaves, flowers and vegetables. Take a look at the head of a cauliflower or a sprig of broccoli and you can see the way in which they repeat the same pattern over and over again on different scales. What an economical plan for the development of complexity.

Size and populations

In any habitat there are always more small creatures than large ones. This is understandable. The large creatures eat the smaller ones and so there is a food chain, which requires an increasing abundance of smaller creatures. And because large creatures eat more, they are more sparsely distributed. They need more territory to hunt in. Bigger animals tend to have fewer young and take longer to rear them to the age when they can breed. This means that they are more susceptible than small animals to extinction by some environmental catastrophe. They will take the longest to regenerate their population. It is no coincidence that so many very large creatures (like mammoths and dinosaurs) have become extinct. On the other hand, the hierarchy of the food chain, which sees large creatures preying upon smaller ones, only leaves an empty niche for the successful evolution of new creatures that are bigger than the biggest that already exist.

One can see that large predatory animals are always in danger of being caught between two opposing constraints. They must not have too high a population density or there will be inadequate food supplies and prey available to support them. But in order to maintain stable population levels they must exist in significant numbers in a particular area. If the scarcity of prey leads to a reduction in their numbers below some critical level then breeding will be less likely, as will the survival of offspring, who will put extra pressure on food reserves. Moreover, the most energy-efficient way for them to feed is to eat animals that are not too much smaller than themselves, otherwise they will waste energy chasing many, small, fast creatures for relatively little return in food. Yet these preferred sources of good are themselves quite rare for the same reasons that their predators are. In a sparse population the chances of extinction by some unexpected change in the environment are also likely to be high. These two competing pressures show why the Earth is not teeming with huge ferocious predators: they just need too much space and too many calories. To continue to thrive they need to enlarge their territory continuously without limit. In practice, the size of land masses, the vagaries of climate and the topography of the land provide obstacles to this expansionism. It is the insects that rule the Earth.

The process of natural selection has given rise to many examples where organisms have evolved much more surface area than would be expected for an object of their volume. One way to achieve this is to develop a wiggly fractal surface, like a natural sponge. In the bathroom you can see that we have learnt this lesson from Nature. Towels retain more water the larger their surface area. So, a typical towel is not smooth. It has hundreds of tiny knots on its surface, which boost the area available for contact with wet surfaces.

The size of brains

If size matters in the living world we can ask whether it has anything to say about future ambitions to design and build artificial forms of intelligence. Are there limits to the size and capabilities of an artificial brain? At first you might think that the bigger the better. But stop and think what computers do, and what *you* do when you think about what *they* do. Each computational step processes information, does work and so produces some waste heat. If we build a larger and larger artificial brain then the volume of its circuitry will grow faster than the area of its surface from which the waste heat can be radiated away. Eventually it will overheat and melt. To overcome this problem we could take a leaf out of the book of Nature and give the computer a crennelated surface so as to boost its area relative to its volume. However, there is a price to pay. In order to keep all the parts of the computer interconnected near that surface we now require a far greater length of circuitry. This means that the computer will operate more slowly, taking longer to send coordinating signals from one part of itself to another. So there seems to be a trade-off between increasing volume, cooling and speed. Perhaps there is some ultimate limit on how big or how powerful a computer can be?

There is a further simple property of things that are large by virtue of containing many components. If one gathers together a collection of *N* objects then they will suffer random fluctuations in their properties that are roughly of a magnitude given by one part in the square root of *N*. Thus we see that small collections are subject to relatively large fluctuations, while very large populations are very stable. This is why an opinion poll needs to sample a large number of people in order to produce an accurate forecast. Our genetic make-up involves a very large number of genes and there are about 10^{29} atoms in a human body. These large numbers, which derive ultimately from our overall size, play an important role (along with genetic error-correction mechanisms) in ensuring that DNA copying errors and other random fluctuations do not reach lethal levels in a very short time. If you want to be complex enough to be 'alive' large size seems to help and possibly even be necessary. It allows the storage of a very large collection of interconnected components and ensures that random statistical fluctuations in their operation are negligible.

Why is the Universe so big?

What about the size of the largest thing we can conceive of—the Universe itself? The visible part of the Universe is about fifteen billion light years in size. Why is it so big? Could we not have lived in a smaller, more economy-sized universe? The answer is surprising. You and I are made out of atoms of chemical elements like carbon, oxygen, nitrogen and phosphorus. These biological elements are produced by the stars. Over billions of years the simple elements of hydrogen and helium are cooked by nuclear reactions within these stars into carbon and heavier elements. When stars exhaust their fuel they undergo supernova explosions which disperse elements like carbon and oxygen into space where they end up in rocks and grains and planets and people. We are indeed made of stardust. But this stellar alchemy, which turns hydrogen into carbon, takes billions of years to complete. So the Universe must be billions of years old if it contains living observers. But why is it so big as well as so old? The Universe is expanding: distant clusters of galaxies are rushing away from each other as if a cosmic explosion occurred long ago. This means that the age of the Universe determines its size. It is billions of light years in size because it has been expanding for billions of years. And so we can now see why our Universe is so big. If it were significantly smaller then it would be too young for stars to have produced the biochemical building blocks of life and we would not be here to wonder about it. The vastness of the Universe is no more, and no less, surprising than the fact of our own existence.

This tells us why we find the Universe to be so old and so large. But just how large is it? It is finite or infinite? In 1915 Albert Einstein published his celebrated new theory of gravitation which extended the classical theory of Newton, which had stood as a monument to human thought since the 17th century. Einstein's theory teaches us that space is curved by the presence of mass and energy within it. Newton's conception of space was that of a flat table top upon which matter moved without ever altering the flatness of the surface. Einstein's conception sees space like a rubber sheet that suffers distortions and indentations when matter moves upon it. Where there are large masses and strong gravitational fields, space is strongly curved, whereas in places where gravity is rather weak the curvature is very small. The most remarkable thing about Einstein's theory of gravity is that it can describe entire universes. In this case the curving of space leads to two quite distinct possibilities. If there is a large enough density of matter in the Universe then it will curve up the shape of space so strongly that it is closed off and finite in volume. This is called a 'closed universe'. But, if the density is lower than some critical value then space can go on forever. This infinitely large space is called an 'open universe'. This state of affairs means that by measuring the density of matter in our universe we can determine whether it is large enough to render the Universe finite. Remarkably, the density lies so close to the critical dividing line that we cannot tell with certainty on which side of it we lie. In fact, the most popular theory of how the Universe behaved during the early moments of its expansion—called the 'inflationary universe theory'—predicts that our universe will deviate from the critical divide by no more than one part in a hundred thousand. It does not tell us on which side of it we lie. Moreover, if we truly lie so close to it then no astronomical observations that we can make will ever be able to tell whether the universe in which we live is infinite or finite in extent.

Just as the population density of creatures on the Earth is limited by their size, so it is with life in the astronomical Universe. But the reasons are different. We have seen that the size of the Universe is inextricably bound up with its age. The long period of time required for the production of biological elements and the subsequent process of molecular and biological evolution ensures that the

Universe is billions of light years in size when life evolves. But the expansion rate of the Universe is dictated by the density of matter in the Universe, because it is the density of matter that determines the strength of gravity, which slows the expanding universe. That density is amazingly small. The critical dividing line separating infinitely expanding universes from those that will eventually recollapse corresponds to an average density in the Universe of only about ten atoms in every cubic metre of space. This is a far emptier vacuum than could ever be produced on Earth by artificial means. Of course, the local density of matter in the solar system is vastly greater than this because it is gathered into dense lumps like planets and people, but if one were to take all the known matter in the Universe and smooth it into a uniform sea of atoms throughout space it would provide no more than between one and ten atoms in every cubic metre. If one thinks about gathering this matter into aggregates one can see how widely separated planets and stars, and thus any civilizations they support, need to be. An average density of ten atoms per cubic metre is the same as just one human being (100 kilograms approximately) in every spherical region of space that is just over a million kilometres in diameter. It is also the same as having just one Earth-sized planet in every region with a diameter of a million billion kilometres and one solar system in every region that is ten times bigger still. So we see that the connection between the size, the age and the density of the Universe guarantees that civilizations in the Universe are likely to be vastly separated from each other.

Last, we come to the least—zero. Does anything have zero size? Our present thinking about the most elementary particles of matter has arrived at a collection of particles that are assumed to possess no internal constituents at all. In practice, this means that when these particles are bombarded by others at very high energies they scatter the incoming particles in a manner that is consistent with their possessing no internal structure at all. They behave like points of zero size, to the accuracy of the experiment. The most powerful scattering experiments of this sort are to be found at the European centre for particle physics (CERN) in Geneva and they show that the most elementary quarks and leptons (a class of particle including the electron, the muon, the tau and all their associated neutrinos) have sizes less than one hundred thousand times smaller than the atomic nucleus. And just as we will be unable to discover if the Universe is infinite, we will be unable to tell if elementary particles truly have zero size. Heisenberg's Uncertainty Principle reveals that more and more energy is necessary to probe structures with smaller and smaller size. An infinite amount of energy would be needed to demonstrate that anything was of zero size.

CHAPTER 6

Are there any laws of physics?

Physics is the interrelationship of everything.

Kirk Dilorenzo

There has grown up, even among many educated people, the view that everything in Nature, every fabrication of its laws, is determined by the local environment in which it was nurtured—that natural selection and the Darwinian revolution have advanced to the boundaries of every scientific discipline Yet, in reality, this is far from the truth. Twentieth-century physicists have discovered that there exist invariant properties of the natural world and its elementary components that render quite inevitable the gross size and structure of almost all its composite objects. The magnitude of bodies like stars, planets, and even people, are neither random nor the result of any progressive selection process but simply manifestations of the different strengths of the various forces of Nature. They are examples of possible equilibrium states between competing forces of attraction and repulsion. A study of how these equilibrium states are set up and how their form is determined, reveals that the structure of the admissible stable states is determined by those parameters we have come to call the fundamental constants of Nature; for example, quantities like the electric charge of the electron, the ratio of the electron and proton masses, the strength of the strong force between nucleons, and so forth.

Suppose we were to commission a survey of all the different types of objects in the Universe from the scale of elementary particles to the largest clusters of galaxies. A picture could be prepared which plotted all the objects according to their mass and their size, or average dimension. *A priori* we might have expected our graph to be covered by points in a fairly haphazard fashion but this is clearly not the case. Some regions of the diagram are heavily and systematically populated, whereas others remain very obviously empty. Our likely

---

A review of *The accidental universe,* by P. C. W. Davies, Cambridge University Press, published 1982

reaction would probably be one of the following three: we could suspect that the points were distributed completely at random—any preference for a particular region of the diagram being purely statistical; all the correlations are real coincidences. Or perhaps we are the victim of a powerful selection effect? Some structures may be unseeable by observers and their existence might explain any areas of significant depopulation in the diagram. Finally, we could try and explain the picture by appeal to stability criteria. The 'rules' of Nature allow only certain types of structure to exist for long periods of time. The populated regions of our diagram are simply those that describe the stable equilibrium states between different natural forces.

This last alternative is the one that successfully and naturally describes the spectrum of objects on view to astronomers. This short book is an expanded version of some earlier review articles and books, well-known to astronomers, which seek to demonstrate how it is possible to deduce the gross characteristics of cosmic objects by a knowledge of dimensional analysis and elementary physical reasoning. The treatment is specifically geared to British undergraduates in that A level physics is used and, unlike in the research literature, SI units prevail. The author first supplies some clear and simple accounts of elementary particle theory, quantum mechanics and relativistic cosmology before moving on to evaluate the necessary scales of structure that emerge as equilibrium states between different forces. These applications form the core of the book but here I found the treatment rather disappointing. Although the exposition remains clear, the choice of material for inclusion and exclusion has not been well made. For instance, although arguments are given to explain the approximate size of asteroids, planets and hydrogen-burning stars there is no discussion of stellar evolution nor a derivation of white dwarf and neutron star sizes. This would have been far more instructive than the more speculative estimates of galactic dimensions which rely on specific assumptions about their mode of formation. It is this part of the discussion that illustrates one side of the book's two-edged title and shows why the gross features that astronomers observe are not accidental. They might have been predicted by someone of sufficient intelligence who knew the laws of physics. There are good reasons why planets and stars and even people come in the proscribed size ranges we see. Why, then, an 'accidental' Universe?

The author introduces the reader to a number of cosmic accidents and a reaction to them called the 'anthropic principle'. The pleasing fact that so many crucial aspects of the Universe are fashioned by (apparently) unchanging properties—the so-called constants of Nature—actually creates our problem. In many cases it has been found that dimensionless combinations of various completely unrelated constants of Nature give pure numbers which have virtually equal values that are extraordinarily large. No explanation for these coincidences exists. A classic example is the rough equality between the ratio of the electric to gravitational forces between two protons and the square root of the number of atoms in the observable Universe—both are roughly equal to ten followed by thirty nine noughts!

In the early 1960s, Robert Dicke pointed out that these coincidences play an important 'humanitarian' role. If they did not exist then neither would we! They codify certain properties of the Universe—like its large size, great age, lack of antimatter, and so forth—which are necessary prerequisites for the evolution and persistence of life as we conceive it. Were the values of the natural constants to differ from what we observe, the Universe really would be *unimaginably* different, for observers like ourselves could not exist. Of all the possible universes that we can conceive of, almost all would be unable to evolve and sustain atomic life. Paradoxically, the uniqueness of our Universe is impressed most forcibly upon us by the fact that we can, in our ignorance, conceive of so many plausible alternatives. The anthropic principle is a label for our recognition of the fact that the Universe allows life to exist because of series of unexplained coincidences (or 'accidents') concerning the magnitudes of its defining constants. Davies gives a summary of various scientists' interpretations of these 'accidents'. Some have tried to invert the logical thread and claim that the concurrence of so many independent accidents provides circumstantial evidence for the strange conclusion that observers are in some sense necessary for the Universe to exist. However, I felt that the author's approach of quoting directly from the writings of the various contributors to this issue was unsuccessful. Several were quoted out of context and the content of their positions diluted to too low a level. The reader who has not met these ideas before needs a more integrated explanation.

The main objection to the anthropic principle is probably the question: Where, or what, are the other defective, life-free worlds that might have been? We can only have grounds for statements of comparative reference regarding the world we see if others really do or could exist. Davies describes some of the available options, notably one which he has written about at a popular level elsewhere: the 'many-worlds' interpretation of quantum mechanics. This requires, for a self-consistent picture of quantum theory (our most successful physical theory), the existence of an infinity of independent realities, through which we weave a path by the continual process of observation. Although most physicists implicitly subscribe to the many-worlds interpretation of quantum mechanics whenever they use the theory, this aspect of it seems to play no role in calculating observable aspects of microphysics and, if pressed, they would probably regard it as a sort of excess metaphysical baggage to be dropped off and picked up again at the door of the laboratory. A more convincing rationale for the ensemble of different possible worlds is necessary if the 'accidental universe' is to be a meaningful and testable statement. Perhaps there is one possibility that has been overlooked and which might, subject to the future course of physics, fit the bill?

If this book had been written twenty, or even ten years ago, it would have described more 'laws' of Nature than today. Human beings have a habit of perceiving in Nature more laws and symmetries than truly exist there. During the past twenty years we have seen a gradual erosion of supposed 'principles' and constants of physics, as Nature has revealed a deep, hidden flexibility of previously unsuspected extent. Many quantities that, traditionally, we believed to be absolutely conserved—parity, charge conjugation, baryon and lepton number—all seem to be violated in elementary particle interactions. The neutrino was always believed to be a massless particle, but recent experiments have provided evidence that it does possess a tiny mass. Similarly, the long-held myth that the proton is an absolutely stable particle is being re-written by recent theoretical arguments and experimental evidence for its instability. Particle physicists have now adopted an extremely revolutionary spirit and it is fashionable to question other long-standing conservation laws and assumptions—is charge conserved, is the photon massless, is the electron stable, is Newton's law of gravity exact at low energy, is the neutron neutral, and so on?

The steady trend from more laws of Nature to less provokes us to ask the overwhelming question: 'Are there any laws of Nature at all?' If the answer is 'no' then the 'accidental' Universe is a particularly appropriate description of the cosmos. It also adds a new and appealing twist to the anthropic principle's dilemma of finding the 'other worlds'. It is possible that the rules we now perceive governing the behaviour of matter and radiation have a purely random origin, and may be an 'illusion', a selection effect of the low energy world we necessarily inhabit. Some preliminary attempts to flesh-out this skeletal idea have shown that even if the underlying symmetry principles of Nature are random—a sort of chaotic combination of all possible symmetries—then it is possible that at low energies the appearance of particular invariances is inevitable under certain circumstances. A form of 'natural' selection may occur wherein, as the temperature of the Universe falls, fewer and fewer of the entire gamut of 'almost symmetries' have a significant impact upon the behaviour of elementary particles, and orderliness arises. Conversely, as the ultimate energy of the Big Bang is approached, this picture would predict chaos. Our low energy world may be necessary for physics as well as physicists.

Let's recall a simpler example of what might be occurring: If you went out into the street and gathered information, say, on the heights of everyone passing-by over a long period of time, you would find that the graph of the frequency of individuals versus height inevitably tended more and more closely to a particular shape. This characteristic 'bell' shape is called the 'normal' or

Gaussian distribution by statisticians. It is ubiquitous in Nature. The Gaussian is characteristic of the frequency distribution of all truly random processes regardless of their specific physical origin. As one goes from one random process to another the resulting Gaussians differ only in their width and the point about which they are centred. A universality of this sort might be associated with the laws of physics if they had a random origin. Suppose that a programme of this sort could be substantiated and provide an explanation of the symmetries of Nature we currently observe; and so, in principle, some of the values of fundamental constants might have a quasi-statistical character. In this case, the anthropic interpretation of Nature might be slightly different. If the laws of Nature are statistical in origin, then again, a *real* ensemble of different possible universes does exist. Our own Universe is one realization of the ensemble. The question now is, are all the features of our Universe stable or generic aspects of the ensemble or are they special? Seen in this light, these 'anarchic' theories are rather attractive to the anthropic interpretation: they allow real, alternative universes as possibilities without incorporating the simultaneous presence of an infinite number of many worlds; they also allow, in principle, a precise mathematical calculation of the probabilities of seeing a particular aspect of the present world and a means of evaluating the statistical significance of any inhabitable universe. In a very general way we can see that the crux of any final analysis of this type, whatever its detailed character, is going to be the temperature of the Universe. Only in a relatively cool Universe, will reliable, invariant laws of Nature be discernible; similarly, however, only in a cool 'accidental' universe can life exist.

This is a stimulating book. Students ought to be encouraged to sample the chapters on order of magnitude analysis since they promote a style of thinking essential to the armoury of the theoretical physicist, and one which often comes as a complete, but welcome, surprise to the new graduate student. Last, but by no means least, the reader will develop a healthy respect for coincidences in physics. Many great advances in our understanding of the Universe have blossomed from the roots of coincidence; for, as Miss Marple once recommended, 'Every coincidence is worth noticing; after all, you can always throw it away later if it is *only* a coincidence'.

CHAPTER 7

Long-distance calls

Several billion trillion tons of superhot exploding hydrogen nuclei rose slowly above the horizon and managed to look small, cold and slightly damp.

Douglas Adams

The Universe is big and old, dark and cold, but its vastness is no accident. Because the Universe is expanding, and billions of years are necessary to produce the building blocks that make the complex molecules needed for life, the Universe must be billions of light years in size if it contains living beings.

Today, an awareness of the vastness of space makes us susceptible to feelings of pessimism and insignificance. But that vastness, and the sheer sparseness of matter, has another curious consequence: extraterrestrials are rare. Our philosophical and religious attitudes, our speculative fiction and fantasy, have all developed with extraterrestrial life as a distant possibility. This may well be a blessing. It ensures that civilisations will evolve independently of each other until they are highly advanced technologically—or, at least, until they have the capability of sending radio signals through space. It also means that their contact with each other is restricted to sending electromagnetic signals at the speed of light. They will not be able to visit, attack, invade, or colonise each other, because of the enormous distances. Direct visits would be limited to tiny robot space-probes, which could reproduce themselves using raw materials available in space. These distances ensure that even radio signals will take aeons to pass between civilisations in neighbouring star systems. No conversations will be possible in real time. The answers to the questions sent by one generation will be received at best by future generations. Conversation will be measured, careful, and ponderous. The insulation provided by interstellar and intergalactic distances protects civilisations from the cultural imperialism of extraterrestrials who are vastly superior. It prevents war, and encourages the art of pure speculation. If you could leapfrog cultural and scientific progression by consulting an oracle who supplies knowledge that would take thousands of years to discover unaided, then the dangers of manipulating things not fully understood would outweigh the benefits. All motivation for human progress might be removed. Fundamental discoveries would be forever out of reach. A decadent, impoverished humanity might result.

The enormity of the Universe, and the vast distances between civilisations within it, has stimulated particular metaphysical attitudes. Many of the enthusiasts who search for signals have argued that signals from more advanced civilisations would enormously benefit humankind. Frank Drake, leader of a long-term search for extraterrestrial intelligence project (SETI), has suggested we are most likely to receive signals from the longest-lived civilisations that have discovered the secret of immortality. One might argue that immortality is not a likely endpoint for the advanced evolution of living beings. Death promotes genetic diversity and periodic extinctions have played a vital role in promoting the spread of life. Frequent, sudden mass extinctions of species in the history of our planet accelerated the evolutionary process. In this respect, a race of immortals would evolve more slowly than mortals.

Immortality does strange things to one's sense of urgency. In his book *Einstein's dreams*, Alan Lightman looks in upon an imaginary world where everyone lives forever. Its society splits into two quite different groups. There are procrastinators who lack all urgency; faced with an eternity ahead of them, there was world enough and time for everything—their motto some word like mañana, but lacking its sense of urgency. Then, there were others who reacted to unlimited time with hyperactivity because they saw the possibility to do everything. But they did not bargain on the dead hand that held back all progress, stopped the completion of any large project, and paralysed society: the voice of experience. When every craftsman's father, and his father, and all his ancestors before him, still lives, then experience ceases to be solely of benefit. There is no end to the hierarchy of consultation, to the wealth of experience, and to the diversity of alternatives. The land of the immortals might well be strewn with unfinished projects, riven by drones and workers with diametrically opposed philosophies of life, and choked by advice.

Death is important in the serial drama of life that evolution directs, both at the genetic and the psychological levels, at least until its positive benefits to the species as a whole can be guaranteed by other means. Human death occurs on a timescale that is short. This has an important impact upon human metaphysical thinking and so dominates the aims and content of most religions. As we have become more sophisticated in curing and preventing disease, the death rate has fallen, and the average human life-span has grown significantly in the richer countries. With this has come a greater fear of death, and a reduced experience of it. There is much speculation about some magic drug or therapy which will isolate the single gene which results in human death by natural causes. By modifying it, some expect that we will be able to prolong the human life-span.

It is unlikely that the evolutionary process would have given rise to organisms with a single weak link dominating the determination of the average life-span. An optimal allocation of resources would see our natural functions wearing out at about the same time, so that no single genetic factor always results in death. Rather, many different malfunctions occur at about the same time. Why allocate scarce resources to develop organs which would work perfectly for 500 years, if other vital organs didn't last for 100 years? Such a budget would fail in competition with one that spread resources evenly around, endowing vital organs with similar life-expectancies. There is a story about Henry Ford which illustrates this strategy. Ford sent a team of investigators to tour the scrapyards of America in search of discarded Model T Fords. He told them he wanted to know which of its mechanical components never failed. They reported failures of just about everything, except the kingpins, which always had years of service left in them. His managers waited to hear how the boss would bring all those failed components up to scratch. Ford announced that in future kingpins on the Model T would be made to a lower specification.

If any extraterrestrial signal were ever to be received, it would have vast philosophical, as well as scientific, significance. The former might well outweigh the latter. If we received a description of some simple piece of physics, it might tell us no facts that we did not already know; but, if it were to use mathematical structures similar to our own, or display similar concepts, like physical constants or laws of Nature, then its philosophical impact would be immense. In mathematics, deep revelations might emerge. If our messages revealed the use of mathematics in a familiar form, with emphasis upon proof, infinite quantities, rather than merely upon computers searching for patterns, then it would shed light upon the nature of mathematics, and its independence of the course of human evolution.

We would expect extraterrestrials to have logic, but would it be our logic? Would they have art or music? Since these are activities that exploit the limited ranges of our human senses, together with aesthetic susceptibilities that have evolved in response to our past environment, we would not expect to find 'art' in an appealing form. Yet, we might expect aesthetic tendencies that reflect the attractiveness of landscapes offering shelter from danger, whilst providing a secure vantage point. An appreciation of these images is an evolutionary by-product of the environments in which most human evolution occurred more than half a million years ago. Such tendencies might be expected to characterise life arising in any environment. Artistic appreciation could even become a predictive activity, with psychologists attempting, on the basis of primary technical or scientific evidence, to predict the aesthetics that might have sprung from them. In language, we might find that the genetic programming that seems to determine human linguistic ability is just one of many ways to achieve a loquacious end; or, we may find our interlocutors displaying grammatical programming similar to our own. Discoveries of this sort would be much more significant than some undiscovered physics that terrestrial physicists could discover for themselves. For what we could learn from a contact about the uniqueness of our concepts, languages, and modes of description, could never be learned without access to an independent civilisation, no matter how far our own studies advanced.

CHAPTER 8

The truth is in the choosing

The universe is built on a plan the profound symmetry of which is somehow present in the inner structure of our intellect.

Paul Valéry

A well-known astronomer once gazed at the journals lining the walls of his university library and whispered: 'It can't all be true you know. There's just not that much truth in the Universe,' 'What is truth?' is a question that has troubled not a few great minds after Pontius Pilate. In seeking an answer to this question, philosophers have traditionally had to grapple with the distinction between the perceived and the unperceived worlds. How large is the difference between the two? Is that difference just the result of an abbreviation of the facts on offer to the mind imposed by the limits of the senses, or is it a serious distortion of some true reality?

Such philosophical conundrums are old but scientific advances highlight them in new ways. The remarkable successes of physical theories based upon extremely abstract mathematics originally developed for other purposes persuades their users that they must be scratching at a bedrock of some true reality. Yet, besides reinforcing the converted in their faith in realism, these successes have produced few methodological principles for epistemologists. More interesting for the theory of knowledge is the impact of evolutionary biology. Immanuel Kant spelt out the limitations upon pure reason imposed by the existence of the mind's innate categories of thought, through which all our sense impressions of true reality are filtered. This anti-Copernican development put each human observer back at the centre of a little world of his or her own making. It left unanswered many questions about the character of those rose-coloured spectacles through which we observe the world. Why are they so similar from one person to the next? Why do they produce an inner picture of the world that is so useful? Could those categories suddenly (or gradually) change? Philosophers took some time to take on board the full implications of

---

A review of *Philosophical Darwinism: on the origin of knowledge by means of natural selection,* by Peter Munz, Routledge, March 1993

the Darwinian revolution. Wittgenstein, who has an unfortunate record of discounting the most important developments, even declared that 'Darwinian theory had nothing to do with philosophy'. In retrospect, the most insightful writer was the Nobel prize-winning biologist Konrad Lorenz, whose 1941 article highlighted the simplifications that might follow from a thoroughgoing philosophical Darwinism. This programme was later dubbed 'evolutionary epistemology' by the psychologist Donald Campbell and can be found in related forms in the early works of Popper that picture the development of scientific theories in the face of falsifying observations as akin to the process of evolution by natural selection. Lorenz's central thesis is very simple. Kant's categories of thought, by which we process information about the world, do not fall ready made from the sky, nor are they correlated from one person to another by some mysterious predestination. They are the results of an evolutionary process that selects those images of the world that are closest to the nature of true reality. Our eyes tell us something about the true nature of light, our ears something of the true nature of sound, because those organs have evolved by pressure of natural selection in response to the true natures of real things called light and sound. Such an argument gives support to a limited realism about our knowledge of the physical world of those areas where an accurate image of reality has selective advantage.

Much of the scholarly debate over this approach to the theory of knowledge has, like Lorenz's original article, appeared in German. Peter Munz has set about systematising and extending the studies of evolutionary epistemology in a thoroughgoing and interesting fashion. Neither a biologist nor a philosopher by training, he brings the critical perspective of the historian to bear upon the complicated growth of ideas in this field and introduces a succession of memorable formulations of the key theses of evolutionary epistemology. Central to his presentation is the picture of living organisms as theories about their environments. They encode information about the real world, which determines their ability to respond more successfully to that environment than their competitors. The stiffer the competition, the better their adaptations need to be and the more accurately they will need to apprehend the nature of the environment. An organism that encodes an incorrect model of that part of the environment necessary for its continued existence will not survive: it becomes a theory of the environment that has been falsified by it.

Munz gives an outline of the main currents of opinion regarding the origins of knowledge in the opening chapters. These are a little turgid, but they are succeeded by a lighter discussion of neural evolution (with much appeal to Edelman's neural Darwinism). His primary goal is to defend a form of 'hypothetical' realism by appeal to the evolutionary process as its guarantor. The most engaging part of his argument is the idea that each organism is an embodied theory about that part of the whole environment with which it interacts during the evolutionary process. It is used repeatedly to render implausible a number of rival theories of knowledge and to disarm some critics of realism. Despite the extensive and sometimes repetitive treatment of these issues, the argument falls short of coming to grips with the fact that the most successful scientific descriptions of the world of elementary particles and astronomical objects are of those parts of Nature that are furthest from our human environment. Either realism is a more powerful presumption than the evolutionary epistemologists would lead us to believe, or the underlying unity of Nature is so strong that our correct apprehension of one part of it goes a long way towards paving the way for a correct image of the whole.

While Munz discusses the biological aspects of adaptation, including the objections of those like Lewontin who are loath to ascribe so much influence to the environment in provoking adaptation, he says virtually nothing about our description of the wider physical world. And in the local biological realm there is scope for a more detailed discussion of which aspects of our categorical apparatus do indeed arise from adaptation rather than from random drift or as neutral evolutionary by-products of selection for other traits. Munz's account impresses upon philosophers the need to incorporate the impact of evolution upon our apprehension of the Universe. It takes us part of the way towards incorporating new discoveries in neurophysiology into the picture and lays out many of the paths that lie untrodden. But there is still much to be said about the evolutionary basis of epistemology before it is adequate to encompass all of our exact knowledge of the Universe. Not before then can it be said that the science of science is a science.

Part 3

Theories of everything—even gravity

Very dangerous things, theories

Dorothy Sayers, *The Unpleasantness at the Bellona Club*

The most dramatic development in fundamental science since 1980 has been the entry into serious scientific discussion of the concept of a 'Theory of Everything'. Its impact has been the greater because of the suddenness of its appearance as much as for the enormity of its agenda. Einstein spent the last 20 years of his life in search of an elusive 'unified field theory' which he believed could tie together the laws of gravity and electromagnetism. Einstein's search was fruitless. There were other forces to be drawn into the unification picture of which too little was known in Einstein's day. Following this failed programme, physicists became more cautious in their aims. All talk of unified theories conjured up pictures of Einstein's failed project and was likely to consign its author to the academic scrap heap. But things changed suddenly and surely. In 1973, particle physicists identified how it was that the forces of Nature worked at high energies. The forces that particles of matter actually felt didn't stay the same as energies and temperatures increased. They changed in such a way that the weak forces got stronger and the strong forces got weaker. This insight paved the way for a deep understanding of how it could be that the forces of Nature, which in our cool, low-energy world billions of years after the inferno of the Big Bang look so different in strength, might be different manifestations of a single underlying force.

In the late 1970s the first 'Grand Unified Theories' (GUTs) were formulated and applied to the study of the early Universe. These theories united the descriptions of the strong, weak, and electromagnetic forces of Nature but were unable to add the gravitational force to the story. All attempts to do so led to a theory that produced infinite answers to questions about measurable quantities. Then, in the early 1980s, a revolution began, which continues to this day. A new sort of theory of matter was found by Michael Green and John Schwarz based upon treating the most elementary entities as strings rather than points.

These tubes of energy possessed a special 'super' symmetry, relating matter and radiation, and were dubbed superstrings. The remarkable thing about superstring theory was that it necessarily included gravity and so offered a fully unified theory of all four forces of Nature. This became known as the 'Theory of Everything', although there appeared to be many different, logically self-consistent, theories of this sort. Recently, it has been shown that these are not really different theories, merely different limiting cases of a single underlying 'Theory of Everything', called M-theory.

This sudden appearance of candidates for a 'Theory of Everything' has led to much confusion about what is meant by 'Everything', especially when these developments have occurred at the same time as other scientists were announcing dramatic developments in our understanding of chaos, unpredictability, and self-organisation. On the one hand, particle physicists seemed to be telling us that the world is very simple, simple enough to be encapsulated within a single law of Nature. On the other hand, everyone else seemed to be saying that the world is amazingly complex, displaying novelty, unpredictability, and intricacy wherever you look. The first chapter in this section, '*Theories of everything*', lays out what we can and cannot expect to learn from a 'Theory of Everything'. The second chapter is about the tantalising question of limits to scientific knowledge of various sorts. It introduces the different sorts of limits we might encounter in our quest to understand the Universe about us: those that are technological and practical, those that are consequences of being the latest editions of an evolutionary process that was not selecting for the ability to do mathematical physics, and those that are deep and fundamental to the nature of things.

The next chapter is a review of a collection of articles written to celebrate the three-hundredth anniversary of the publication of Newton's great book, the *Principia*. Newton's theory of gravity held sway for more than two hundred years until Einstein showed that it was just a limiting case of a more general theory. We are still waiting to discover the yet more general theory (of everything?) of which Einstein's great theory is, in turn, just a limiting case. Perhaps we are closer to finding it than we could ever have dared hope.

Finally, there is a review of a recent book by the biologist E. O. Wilson that foresees a future in which all branches of scientific study will be coordinated. Wilson's state of 'consilience' will not stop there though: its reach will encompass many of the creative acts, religious inquiry, and psychology. Despite the overly reductionist ambitions voiced therein, Wilson's book is appealing because of his passionate appeal for an environmental appraisal that deals properly with full-cost economic accounting of the negative consequences of human actions. It is a challenge to politicians at the end of the millenium.

CHAPTER 9

Theories of everything

We are all in the gutter; but some of us are trying to get published in Gutt. Rev. Lett.

Robert McLachlan

Introduction

Despite the topicality of 'Theories of Everything'[1] in the literature of science and its popular chronicles, they are, at their root, new editions of something very old indeed. If we cast our eye over a range of ancient mythological accounts of the world we soon find that we have before us the first 'theories of everything'. Their authors composed an elaborate story in which there was a place for everything and everything had its place.[2] These were not, in any modern sense, scientific theories about the world, but tapestries within which the known and the unknown could be interwoven to produce a single picture with a 'meaning' in which the authors could place themselves with a confidence borne of their interpretation of the world around them. In time, as more things were discovered and added to the story, so it became increasingly contrived and complicated. Moreover, whilst these accounts aimed at great breadth when assimilating perceived truths about the world into a single coherent whole, they were totally lacking in depth, namely in the ability to extract more from the story than was put into it in the first place. Modern scientific theories about the world place great emphasis upon depth—the ability to predict new things and explain phenomena not incorporated in the specification of the theory in the first place. For example, an explanation for the living world that maintained it was created, ready-made, just hundreds of years ago but accompanied by a fossil record with the appearance of billions of years of antiquity, certainly has breadth; but it is shallow. Experience teaches us that it is most efficient to begin with a theory that is narrow but deep and then seek to extend it into a description that is both broad and deep.[1] If we begin with a picture that is broad and shallow we have insufficient guidance to graduate to a correct description that is broad and deep.

In recent years there has been renewed interest in the possibility of a 'Theory of Everything'.[1] In what follows we shall see what is meant by a 'Theory of Everything' and how, while it may be necessary for our description of the Universe and its contents, it is far from sufficient to complete our understanding. We cannot 'reduce' everything we see to a 'Theory of Everything' of the particle physicists' sort. Other factors must enter to complete the scientific description of the Universe. One of the lessons that will emerge from our account is the extent to which it is dangerous to draw general conclusions about 'science' or the 'scientific method' in the course of discussing an issue like reductionism or the relative merits of religion and science. Local sciences, like biology or chemistry, are quite different to astronomy or particle physics. In the local sciences one can gather virtually any data one likes, perform any experiment, and (most important of all) one has control over possible sources of bias introduced by the experimental set-up or the process of gathering observations. Experiments can be repeated in different ways. In astronomy this is no longer the case: we cannot experiment with the Universe; we just have to take what is on offer. What we see is inevitably biased by our existence and what is needed for objects to be visible: intrinsically bright objects are therefore always over-represented in our surveys. Likewise, in high-energy particle physics there are serious limitations imposed upon our ability to experiment. We cannot achieve the very high energies that are required to unlock many of the secrets of the elementary particle world by direct experiment. The philosophy of science has said a lot about scientific method under the assumption of an ideal environment in which any desired experiment is possible. It has not, to my knowledge, addressed the situation beset by limited experimental possibilities with the same enthusiasm.

Order out of chaos

Suppose you encounter two sequences of digits. The first has the form

. . . 0010010010010010 01 . . .

whilst the second has the form

. . . 010010110101111010010 . . .

Now you are asked if these sequences are random or ordered. Clearly, the first appears to be ordered and the reason you say this is because it is possible to 'see' a pattern in it; that is, we can replace the sequence by a rule that allows us to remember it or convey it to others without simply listing its contents. Thus, we will call a sequence non-random if it can be abbreviated by a formula or a rule shorter than itself. If this is so we say that it is *compressible*.[3] On the other hand, if, as appears to be the case for the second sequence (which was generated by tossing a coin), there is no abbreviation or formula that can capture its information content then we say that it is *incompressible*. If we want to tell our friends about the incompressible sequence then we simply have to list it in full. There is no encapsulation of its information content shorter than itself.

This simple idea allows us to draw some lessons about the scientific search for a 'Theory of Everything'. We might define science to be the search for compressions. We observe the world in all possible ways and gather facts about it; but this is not science. We are not content, like crazed historians, simply to gather up a record of everything that has ever happened. Instead, we look for patterns in those facts, compressions of the information on offer, and these patterns we have come to call the laws of Nature. The search for a 'Theory of Everything' is the quest for an ultimate compression of the world. Interestingly, Chaitin's proof of Gödel's incompleteness theorem[4] using the concepts of complexity and compression reveals that Gödel's theorem is equivalent to the fact that one cannot prove a sequence to be incompressible. We can never prove a compression to be the ultimate one; there might be a yet deeper and simpler unification waiting to be found.

There is a further point that we might raise regarding the quest for a 'Theory of Everything'—if it exists. We might wonder whether such a theory is buried deep (perhaps infinitely deep) in the nature of the Universe or whether it lies rather shallow. One suspects it to lie deep in the structure of things and so it might appear a most anti-Copernican over-confidence to expect that we would be able to fathom it after just a few hundred years of serious study of the laws of Nature, aided by limited observations of the world by relatively few individuals. There appears to be no good evolutionary reason why our intellectual capabilities need be so great as to unravel the ultimate laws of Nature unless those ultimate laws are simply a vast elaboration of very simple principles, like counting or comparing,[5] which are employed in local laws. Of course, the unlikelihood of our success is no reason not to try. It is just that we should not have unrealistic expectations about the chances of success and the magnitude of the task.

Laws of Nature

Our discussion of the compressibility of sequences has taught us that pattern, or symmetry, is equivalent to laws or rules of change.[6] Classical laws of change, like Newton's law of motion and momentum conservation are equivalent to the invariance of some quantity or pattern. These equivalences only became known long after the formulation of the laws of motion that governed the allowed changes. This strikes a chord with the traditional Platonic tradition, which places emphasis upon the unchanging, atemporal aspects of the world as the key to its fundamental structures. These timeless attributes, or 'forms' as Plato called them, seem to have emerged with the passage of time as the laws of Nature, or the invariances and conserved quantities (like energy and momentum) of modern physics.

Since 1973 this focus upon symmetry has taken centre stage in the study of elementary-particle physics and the laws governing the fundamental interactions of Nature. Symmetry is now taken as the primary guide into the structure of the elementary-particle world and the laws of change are derived from the requirement that particular symmetries, often of a highly abstract character, be preserved. Such theories are called 'gauge theories'.[7] The currently successful theories of the known forces of Nature—the electromagnetic, weak, strong, and gravitational forces—are all gauge theories. These theories require the existence of the forces they govern in order to preserve the invariances upon which they are based. They are also able to dictate the character of the elementary particles of matter that they govern. In these respects they differ from the classical laws of Newton, which, since they governed the motions of all particles, could say nothing about the properties of those particles. The reason for this added dimension is that the elementary-particle world that the gauge theories rule, in contrast to the macroscopic world, is populated by collections of identical particles ('once you've seen one electron you've seen 'em all').

The use of symmetry in this powerful way enables entire systems of natural laws to be derived from the requirement that a certain abstract pattern be invariant in the Universe. Subsequently, the predictions of these systems of laws can be compared with the actual course of Nature. This is the opposite route to that which might have been followed a century ago. Then, the observational study of events would lead to systems of mathematical equations giving the laws of change; afterwards, the fact that they are equivalent to some global or local invariance might be recognised.

This generation of theories for each of the separate interactions of Nature has motivated the search for a unification of those theories into more comprehensive editions based upon larger symmetries within which the smaller symmetries respected by the individual forces of Nature might be accommodated in an interlocking fashion that places some new constraint upon their allowed forms. So far this strategy has resulted in a successful, experimentally tested, unification of the electromagnetic and weak interactions, and a number of purely theoretical proposals for a further unification with the strong interaction ('grand unification'), and ultimately a four-fold unification with the gravitational force to produce a so called 'Theory of Everything' or 'TOE'.[1] The pattern of unification that has occurred over the last three-hundred years is shown in Figure 9.1.

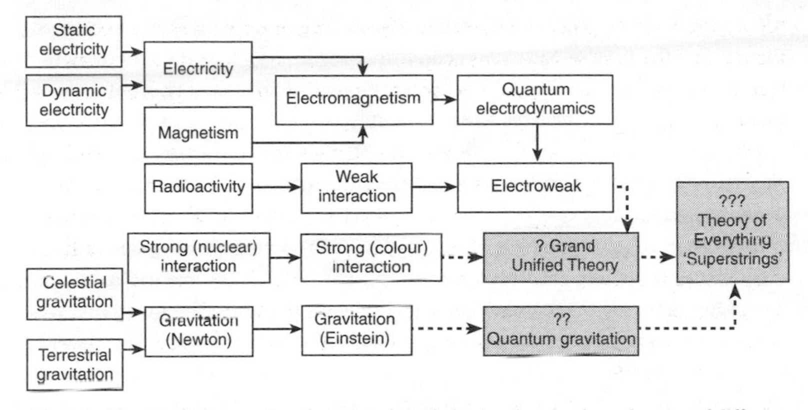

Fig. 9.1. The development of various theories of physics showing how theories of different forces of Nature have become unified and refined. The dotted lines represent unifications that have so far been made only theoretically and have yet to be confirmed by experimental evidence.

The current favoured candidate for a TOE is a superstring theory first developed by Michael Green and John Schwartz.[8] Elementary descriptions of its workings can be found elsewhere.[1,9] Suffice it to say that the enormous interest that these theories have attracted can be attributed to the fact that they revealed that the requirement of logical self-consistency—suspected of being a rather weak constraint upon a TOE—turned out to be enormously restrictive. At first it was believed that it narrowed the alternatives down to just two possible symmetries underlying the TOE. Subsequently, the situation has been found to be rather more complicated than first imagined and superstring theories have been found to require new types of mathematics for their elucidation.

The important lesson to be learned from this part of our discussion is that 'Theories of Everything', as currently conceived, are simply attempts to encapsulate all the laws governing the fundamental natural forces within a single law of Nature derived from the preservation of a single overarching symmetry. We might add that, at present, four fundamental forces are known, of which the weakest is gravitation. There might exist other, far weaker forces of Nature that are too weak for us to detect (perhaps ever) but whose discovery is necessary if we are to identify that single 'Theory of Everything'.

Outcomes and broken symmetries

If you were to engage a particle physicist in a conversation about the nature of the world he might soon be regaling you with a story about how simple and symmetrical the world really is if only you look at things in the right way. But when you return to contemplate the real world you know that it is far from simple. For the psychologist, the economist, the botanist, or the zoologist the world is quite the opposite. It is a higgledy-piggledy of complex events whose nature owes more to their persistence or stability over time than any mysterious attraction for symmetry or simplicity. So who is right? Is the world really simple, as the physicist said, or is it complex as everyone else seems to think?[10]

The answer to this question reveals one of the deep subtleties of the Universe's structure. When we look around us we do not observe the laws of Nature; rather, we see the outcomes of those laws. There is a world of difference. Outcomes are much more complicated than the underlying laws because they do not have to respect the symmetries displayed by the laws. By this we mean that it is possible to have a world that displays complicated asymmetrical structures (like ourselves) and yet is governed by very simple, symmetrical laws. Consider the following simple example. Suppose I balance a ball at the apex of a cone. If I were to release the ball then the law of gravitation will determine its subsequent motion. But gravity has no preference for any particular direction in the Universe; it is entirely democratic in that respect. Yet, when I release the ball it will always fall in some particular direction, either because it was given a little push in one direction or as a result of quantum fluctuations, which do not permit an unstable equilibrium state to persist. Thus, in the outcome of the ball falling down, the directional symmetry of the law of gravity is broken. Take another example. You and I are at this moment situated at particular places in the Universe, despite the fact that the laws of Nature display no preference for any one place in the Universe over any other. We are both (very complicated) outcomes of the laws of Nature that break their underlying symmetries with respect to positions in space. This teaches us why science is often so difficult. When we observe the world we see only the broken symmetries manifested through the outcomes of the laws of Nature and from them we must work backwards to unmask the hidden symmetries that characterize the laws.

We can now understand the answers that we obtained from the different scientists. The particle physicist works closest to the laws of Nature and so is impressed by their simplicity and symmetry. But the biologist, or the meteorologist, is occupied with the study of the complex outcomes of the laws, rather than with the laws themselves and so is impressed by the complexities of Nature rather than by its laws. This dichotomy is displayed in Figure 9.2.

The left-hand column represents the development of the Platonic perspective on the world, with its emphasis upon the unchanging elements behind things —laws, conserved quantities, symmetries—whereas the right-hand column, with its stress upon time and change and the concatenation of complex happenings, is the fulfilment of the Aristotelian approach to understanding the world. Until

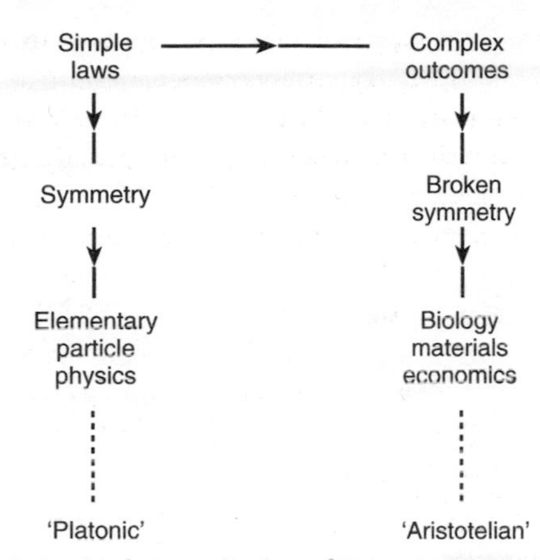

Fig. 9.2. The inter-relationship between the laws of Nature and the outcomes of those laws, together with the scientific disciplines that focus primarily upon them. The left-hand thread from the laws of nature is in the Platonic tradition, while the right-hand line focuses upon temporal development and complicated outcomes and is associated with the Aristotelian perspective upon the world.

rather recently, physicists have focused almost exclusively upon the study of the laws rather than the complex outcomes. This is not surprising, for the study of the outcomes is a far more difficult problem, which requires the existence of powerful interactive computers with good graphics for its full implementation. It is no coincidence that the study of complexity and chaos[11] in that world of outcomes has gone hand-in-hand with the growing power and availability of low-cost personal computers.

We see that the structure of the world around us cannot be explained by the laws of Nature alone. The broken symmetries around us may not allow us to deduce the underlying laws, and a knowledge of those laws may not allow us to deduce all the permitted outcomes. Indeed, the latter state of affairs is not uncommon in fundamental physics and is displayed by the current state of superstring theories. Theoretical physicists believe they have the laws (i.e. the mathematical equations) but they are unable to deduce the outcomes of those laws (i.e. find the solutions to the equations). Thus, we see that while 'Theories of Everything' may be necessary to understand the world we see around us, they are far from sufficient.

Of the complex outcomes of the laws of Nature, much the most interesting are those that display what has become known as 'organised complexity'. A selection of these were displayed in Chapter 4 (page 21) in terms of their size (gauged by information storage capacity) versus their ability to process information (change one list of numbers into another list).

Increasingly complex entities arise as we proceed up the diagonal, where increasing information storage capability grows hand-in-hand with the ability to transform that information into new forms. These complex systems are typified by the presence of feedback, self-organisation, and emergent behaviour. There might be no limit to the complexity of the entities that can exist further and further up the diagonal. Thus, for example, a complex phenomenon like high-temperature superconductivity,[13] which relies upon a very particular mixture of materials being brought together under special conditions, might never have been manifested in the Universe before the right mixtures were made on Earth in 1987. It is most unlikely that these mixtures occur naturally in the Universe and so one variety of complexity, called 'high-temperature superconductivity', relies upon another variety of complexity, that we call 'intelligence', to act as a catalyst for its creation. Moreover, we might speculate that there exist new types of 'law' or 'principle' that govern the existence and evolution of complexity defined in some abstract sense.[14] These rules might be quite different from the laws of the particle physicist and might not be based upon symmetry and invariance but instead upon principles of logic and information processing.

A defining characteristic of the structures displayed on page 21 is that they are more than the sum of their parts.[15] It is the circuit diagram or the neural network that is responsible for the complexity of their behaviour. The laws of electromagnetism alone are insufficient to explain the workings of a brain. We need to know how it is wired up. No 'Theory of Everything' that the particle physicists supply us with will shed this light upon the workings of the human brain or the nervous system of an elephant.

So far we have discussed the outcomes of the laws of Nature using rather straightforward examples but we shall find that there are some aspects of the Universe, which were once treated as unchanging parts of its constitution, that have gradually begun to appear more and more like asymmetrical outcomes; indeed, the entire Universe may be one such asymmetrical outcome of an underlying law whose ultimate symmetry is now hidden from us.

What are the ultimate rules of the game?

Physicists tend to believe that a 'Theory of Everything' will be some set of equations governing entities like points or strings that are equivalent to the preservation of some symmetry that underlies things. This is an extrapolation of the direction in which particle physics has been moving for some time. A key assumption of such a picture is that it regards the laws of physics as being the bottom line and these laws govern a world of point particles or strings (or other exotica) that is a continuum. Another possibility is that the Universe is not, as its root, a great symmetry but a computation. The ultimate laws of Nature may be akin to software running upon the hardware provided by elementary particles and energy. The laws of physics might then be derived from some more basic principles governing computation and logic. This view might have radical consequences for our appreciation of the subtlety of Nature because it seems to require that the world is discontinuous, like a computation, rather than a continuum. This makes the Universe a much more complicated place. If we count the number of discontinuous changes that can exist we find that there are infinitely many more of them than there are continuous changes. By regarding the bedrock structure of the Universe as a continuum we may not just be making an approximation but an infinite simplification.

References

1 J.D. Barrow, *Theories of everything: the quest for ultimate explanation*, Oxford University Press, 1991.

2 C.H. Long, *Alpha: the myths of creation*, G. Braziller, 1963.

3 G. Chaitin, *Algorithmic information theory*, Cambridge University Press, 1987.

4 G. Chaitin, 'Randomness in arithmetic', *Scientific American* July, p. 80, 1980.

5 J.D. Barrow, *Pi in the sky: counting, thinking of being*, Oxford University Press, 1992.

6 J.D. Barrow, *The world within the world*, Oxford University Press, 1988; R. Feynman, *The character of physical law*, MIT Press, 1965.

7 H. Pagels, *Perfect symmetry*, Michael Joseph, 1985; A. Zee, *Fearful symmetry: the search for beauty in modern physics*, MacMillan, 1986; S. Weinberg, *The discovery of sub-atomic particles*, W.H. Freeman, 1983.

8 M. Green, J. Schwarz, and E. Witten, *Superstring theory*, (2 vols), Cambridge University Press, 1987.

9 M. Green, 'Superstrings', *Scientific American*, September, p. 48, 1986; D. Bailin, 'Why superstrings?', *Contemp. Phys.* **30**, 237 (1989); P.C.W. Davies and J.R. Brown, (eds), *Superstrings: a theory of everything?*, Cambridge University Press, 1988.

10 J.D. Barrow, 'Platonic relationships in the Universe', *New Scientist*, 20th April, p. 40, 1991.

11 J. Gleick, *Chaos: making a new science*, Wiking, 1987; I. Stewart, *Does God play dice: the mathematics of chaos*, Blackwell, 1989; B. Mandelbrot, *The fractal geometry of nature*, W.H. Freeman, 1982.

12 H. Moravec, *Mind children*, Harvard University Press, 1988.

13 C. Gough, 'Challenges of high-T_c', *Physics World*, December, p. 26, 1991.

14 S. Lloyd and H. Pagels, 'Complexity as thermodynamic depth', *Annals of Physics (New York)* **188**, 186 (1988).

15 P.C.W. Davies, *The cosmic blueprint*, Heinemann, 1987.

CHAPTER 10

Limits of science

Prediction is difficult, especially about the future.

Niels Bohr

Introduction

An apocryphal story is often told of the patent office whose director made an application for his office to be closed at the end of the nineteenth century because all important discoveries had already been made.[1] Such a fable nicely illustrates the air of overconfidence in human capabilities that seems to attend the end of every century. In science this confidence often manifests itself in expectations that our study of some branch of nature will soon be completed. Typically, there is great confidence in the scope of a successful line of enquiry, so much so that it is expected to solve all problems within its encompass. But the quest to complete its agenda often uncovers a fundamental barrier to its completion—an impossibility theorem.

I want to draw attention to some of the barriers to scientific progress that may be encountered in the future—some of which have been encountered already. They are interesting, not because of some satisfaction at seeing science circumscribed, but because they involve key ideas in science that are as important as new discoveries. As a particular example, I shall focus the discussion primarily upon the search for a 'Theory of Everything'—by which particle physicists mean simply a unified description of the laws governing the fundamental forces of Nature.[2] I want to consider four general types of limitation that could prevent the completion of our investigations into the form of a 'Theory of Everything'.

Existential limits

The first question we should pose is whether such a 'Theory of Everything' exists at all. It is possible that some of our research programmes are directed at discovering theoretical structures that do not exist. At root this is just the old philosophical problem of distinguishing between knowledge of the world and knowledge of our mental models of the world. Limitations upon our abilities to understand fully the latter might be best interpreted as the 'limits of scientists' rather than the 'limits of science'. Bearing this distinction in mind we should be sensitive to the sources of some of our favourite scientific concepts.

The notion that it is more desirable to seek a description of the Universe in terms of a single force law, rather than in terms of 2, 3, or 4 forces, is, at its root, like the entire concept of a law of Nature, a perspective that has clearly definable religious origins.[2] Although for a number of years there has been growing interest in the structure of superstring and conformal field theories as candidates for a 'Theory of Everything',[3] these have proven too difficult to solve. Hence, there are as yet no predictions, observations, or tests to decide whether or not this line of inquiry is in accord with the structure of the world, rather than merely a new branch of pure mathematics.

There have been some studies of particle physics theories in which there is no unification of the fundamental forces of Nature at very high energies. So-called 'chaotic gauge theories'[4] assumed that there were no symmetries at all at very high energy; these emerged only in the low-temperature limit of the theory, a limit that necessarily describes the world in which atom-based life-forms live. These ideas have not yet been thoroughly investigated, but they are similar in spirit to some of the recent quantum cosmological studies on the nature of time. They argue that time is a concept that only emerges in the low temperature limit of a quantum cosmological theory, when the Universe has expanded to a large size compared with the Planck length scale of 10^{-33} cm.[5]

Even if one concedes that there is a 'Theory of Everything' to be found, it should not be assumed that it is logically unique. When the study of superstring theories began in the early 1980s, it appeared that there might be just two of these theories to choose from.[6] At first, in these theories, logical self-consistency appeared to be a much more powerful restriction than had previously been imagined. As investigations have proceeded, a new perspective has emerged. There appeared to be many possible string theories. But it may ultimately transpire that these theories are not as different as they at first appear, many of them may turn out to be just different ways of representing the same underlying theory. If they are distinct then the message may be that there exist many different, self-consistent 'Theories of Everything'. Some of them may contain the four forces of Nature that we observe, while others may lack some of these forces—or contain additional forces.

While we know that we inhabit a world described by a system of forces that permits the existence and persistence of stable complex systems—of which DNA-based 'life' is a particular example—if other systems of forces also permitted complex life to exist, then we would need to explain why one logically consistent 'Theory of Everything' was chosen rather than another.

Conceptual limits

If a deep Theory of Everything does exist, then how confident should we be about our ability to comprehend it? This depends upon how deep a structure it is. We could imagine an infinitely deep sequence of structures that we could only ever partially fathom. Alternatively, the 'Theory of Everything' may lie only slightly below the surface of appearances and be well within our grasp to comprehend. It does not follow that the most fundamental physical laws need be the deepest and most logically complicated aspects of the universal structure.

In practice, we have learned that the outcomes of the laws of Nature are invariably far more complicated than the laws themselves because they do not have to possess the same symmetries as the laws themselves.[2] However, we must appreciate that the human brain has evolved its repertoire of conceptual and analytical abilities in response to the specific challenges posed by the tropical savannah environments in which our ancient ancestors evolved[7] over half a million years ago. There would seem to be no evolutionary need for an ability to understand elementary-particle physics, black holes, and the ultimate laws of Nature. Indeed, it is not even clear that something as simple as rationality was selected for in the evolutionary process. This pessimistic expectation could be avoided if it is true that the laws of nature can be understood in full detail by a combination of very elementary concepts—like those of counting, cause-and-effect, symmetry, and so on—conceptions that do seem to have adaptive survival value. In that case, our scientific ability should be seen as a by-product of adaptations to environmental challenges that may no longer exist, or which are dealt with in other ways following the emergence of consciousness. Moreover, much of our most elementary intuition for number and quantity may be a by-product of our linguistic instincts. It might even be that the structure of counting systems in many primitive and ancient cultures, together with notions like the place/value notation,[8] derive from our complex genetic programming for language acquisition. Our linguistic abilities are far more impressive than our mathematical abilities, both in their complexity and their universality amongst humans of all races.

We might ask whether a 'Theory of Everything' will be mathematical. All our scientific studies of the Universe assume that it is well described by mathematical structures. Indeed, some would say much more: that the Universe is a mathematical structure.[8] Is this really a presumption? We can think of mathematics as being the description (or the collection) of all possible patterns. Some of these patterns have physical manifestations, while others are more abstract. Defined in this way, we can see that the existence of mathematics is inevitable in universes that possess structure and pattern of any sort. In particular, if life exists then pattern must exist, and so must mathematics. There is at present no reason to believe that there exists any type of structure that could not be described by mathematics. But this does not mean that the application of mathematics to all structures will prove fruitful. Indeed, the point is tautological: given another type of description, this would simply be added to the body of what we call mathematics.

Let us return to the issue of our evolutionary development, and the debt we owe to it. One way to look at the evolutionary process is as a means by which complex ('living') things produce internal models, or bodily representations, of parts of the environment. Some aspects of the astronomical Universe—like its vast age and size—are necessary prerequisites for the existence of life in the Universe. Billions of years are required for nuclear reactions within stars to produce elements heavier than helium, which form the building blocks of complexity. Because the Universe is expanding, it must also be at least billions of light years in size. We can see how these necessary conditions for the existence of life in the Universe also play a role in fashioning the view of the Universe that any conscious life-forms will develop.

In our own case, the fact that the Universe is so big and old has influenced our religious and metaphysical thinking in countless ways. The fact that the Universe is expanding also ensures that the night sky is dark and patterned with groups of stars, and this feature of our environment has played a significant role in the development of many mythologies and attitudes towards the unknown.[7] It has also played an important role in the course of evolution by natural selection on the Earth's surface. Many of the general ideas that we have about the nature of the Universe, and its origins, have their basis in these metaphysical attitudes, which have been subtly shaped within our minds by the cosmic environment about us.

Technological limits

Even if we were able to conceive of and formulate a 'Theory of Everything'—perhaps as a result of principles of logical self-consistency and completeness alone—we would be faced with an even more formidable task: that of testing it by experiment. There is no reason why the Universe should have been constructed for our convenience. The decisive features predicted by such theories might well lie beyond the reach of our technology. We appreciate that this is no idle speculation. In recent years the US Congress has cancelled the Superconducting Super Collider project on the grounds of its cost, and the future of experimental particle physics now rests in the hands of the Large Hadron Collider project planned for CERN. These projects seek to increase collider energies far into the TeV (tera electron volt, 10^{12} eV) range to search for evidence of supersymmetry, the top quark, and the Higgs boson—all pieces in the standard model of elementary particle physics.[9] However, even the energies expected in the LHC fall short by a factor of about one million billion of those required to test directly the pattern of fourfold unification proposed by a 'Theory of Everything'. Restrictions of economics and engineering, the pressing nature of other more fruitful and vital lines of scientific inquiry, and the collective wishes of voters in the large democracies doom such direct probes of the ultra-high energy world.

Unfortunately for us, the most interesting and fundamental aspects of the laws of Nature are intricately disguised and hidden by symmetry-breaking processes that occur at energies far in excess of those experienced in our temperate, terrestrial environment. The laws of Nature are only expected to display their true simplicity at unattainably high energies. According to some superstring theories, the situation is even worse: the laws of Nature are predicted only to exhibit that simplicity and symmetry in more space dimensions than the three that we inhabit. The Universe may have nine or twenty-five spatial dimensions, but only three of those dimensions are now large and visible; that rest are confined to tiny length scales, far too small for us to scrutinize directly.[3,10] Our best hope of an observational probe of them is if there are processes that leave some small, but measurable, traces which trickle down to very low energies.

Alternatively, particle physicists look increasingly at astronomical environments to produce the extreme conditions needed to manifest the subtle consequences of their theories. In theory, the expanding universe experienced arbitrarily high energies during the first moments of its expansion. The nature of a 'Theory of Everything' would have influenced the character of those first moments and may have imprinted features upon the Universe that are observable today.[11] However, the most promising theory of the behaviour of the very early Universe—the inflationary universe—dashes such hopes. Cosmological inflation requires the Universe to have undergone a brief period of accelerated expansion during its early expansion history.[12] This has the advantage of endowing the Universe with various structural features that we can test today. However, it has the negative effect of erasing all information about the state of the Universe before inflation occurred. Inflation wipes the slate clean of all the information the Universe carries about the Planck epoch when the 'Theory of Everything' would have its principal effects upon the structure of the Universe.[11]

Limits upon the attainment of high energies are not the only technological restrictions upon the range of experimental science. In astronomy, we appreciate that the bulk of the Universe exists in some dark form whose existence is known to us only through its gravitational effects.[13] Some of this material is undoubtedly in the form of very faint objects and dead stars, but the majority is suspected to reside in populations of very weakly interacting particles. Since there are limits on our abilities to detect very faint objects and weakly interacting particles in the midst of other brighter objects, and strongly interacting particles, there may be a technological limit upon the extent to which we can determine the identity of the material content of the Universe. Likewise, the search for gravitational waves might, if we are unluckily situated in the Universe, require huge laser interferometers or space-based probes in order to be successful, and might therefore, ultimately, be limited by costs and engineering capabilities in just the same way that high-energy physics has proved to be.

At present we know of four forces of nature, and attempts to create unified theories of physics that join them into a single superforce described by a 'Theory of Everything' assume that only these four forces exist. But other forces of Nature could exist, so weak that their effects are totally insignificant both locally and astronomically, but whose presence in the 'Theory of Everything' is crucial for determining its identity. There is no reason why the forces of Nature should all have strengths that are such that we can detect them with our present technology.

It is interesting to recall how accidents of our own location in the Universe have made the growth of technical science possible in many areas. Even if some hypothetical extraterrestrial life-form required high intelligence to survive in its local environment, it should not be assumed that this means that they will have highly developed scientific knowledge in all areas.

For example, it is an accident of geology that our planet is well endowed with accessible surface metals. Without them, no technology would have developed. The existence of the Earth's magnetic field, together with the presence in the Earth's crust of magnetic and radioactive materials, has led to our understanding of these forces of nature. Similarly, an accident of meteorology has saved the Earth from a permanent sky-covering of cloud, and enabled us to develop astronomical understanding. And, even with cloud-free skies, we have benefited from an accident of astronomy: our particular location in the disk of the Milky Way could easily have been shrouded by dust in all directions (rather than just in the plane of the Milky Way), so inhibiting the development of optical astronomy. Our ability to test Einstein's theory of gravitation hinged originally upon two coincidences about our solar system. The fact that the full Moon and the Sun have the same apparent size in the sky (despite being very different in real size, and in distance from us) means that complete eclipses of the Sun can be seen from Earth. This enabled the light-bending predictions of Einstein's general theory of relativity to be tested in the first half of this century. Similarly, the presence of a planet with an orbit as close to the Sun as Mercury's enabled the predictions of perihelion precession by planetary orbits to be checked. Without these fortunate circumstances scientists would have been left with a largely untestable theory. These examples are given simply to make the point that scientific progress is not necessarily an inevitable march of progress that will occur in any civilization that is 'advanced' by some criterion. There may be accidents of environment that prevent the development of science in some directions, while facilitating it in others.

Fundamental limits

One hallmark of the progress of a mature science is that it eventually begins to appreciate its own boundaries. In the present century we have seen many examples of this. In an attempt to extend a theory in new ways some form of 'impossibility' theorem has often been discovered—a proof using the assumptions of the theory that certain things cannot be done or certain questions cannot be answered. The most famous examples are Heisenberg's Uncertainty Principle in physics, Einstein's speed-of-light limit on signalling velocities, Gödel's incompleteness theorem in mathematical logic, Arrow's theorem in economics, Turing uncomputability, the intractability of NP-complete problems, like the 'travelling salesman problem', and Chaitin's theorem about the unprovability of algorithmic randomness. It may transpire that these impossibility results, together with many others that are suggested by studies of space–time singularities, space–time horizons, and information theory, may place real restrictions on our ability to frame or test a 'Theory of Everything'.

We know already that the finite velocity of light ensures that we have a visual horizon (about fifteen billion light years away) in the Universe, beyond which light has not had time to reach us since the expansion of the Universe began. Thus, we are always prevented from ascertaining the structure of the entire Universe (which may be infinite in extent). Astronomers are confined to studying a finite portion of it, called the visible universe. It may be that the visible universe does not contain enough information to characterize the laws of physics completely. Certainly, it does not carry enough information to determine the nature of any initial state for the whole Universe without the introduction of unverifiable, and necessarily highly speculative 'principles' to which the initial state is believed to adhere.[11,14] Many such principles have been proposed—the 'no boundary condition' of Hartle and Hawking,[15] the minimum gravitational entropy condition of Penrose,[16] and the out-going wave condition of Vilenkin[17] are well known examples. Global principles of this sort will all provide quantum-averaged specifications of the entire cosmological initial state, but our visible universe today is the expanded image (possibly reprocessed by inflation) of a tiny part of that initial state, where conditions may deviate from the average in some way, if only because we know that they happened to satisfy the stringent conditions necessary for the eventual evolution of living complexity.

As yet, Gödelian incompleteness has not made any restriction upon our physical understanding of the Universe, although it has the scope to do so because it has been shown by Chaitin[8,18] that it can be recast as a statement that sequences cannot be proved to be random (a Gödel undecidable statement might be just the one needed to characterize the order in a sequence). Gödel's theorem requires that logical systems large enough to contain the whole of arithmetic are either inconsistent or incomplete. Now, although modern physicists use the whole of arithmetic (and much more besides) to describe the physical universe, we cannot conclude from this that there will exist some undecidable proposition about the Universe. The description of the laws of Nature may require only the decidable part of mathematics. Alternatively, it may not be necessary to use a mathematical structure as rich as the whole of arithmetic to characterize the Universe. The fact that we do so may simply be because it is quick and convenient, or because of a sequence of historical accidents that have bequeathed a particular mathematical system. If one uses a smaller logical system than arithmetic (like Euclidean geometry, or Presburger arithmetic, with only + and − operations) then there is no Gödel incompleteness: in these simpler axiomatic systems all statements can be demonstrated to be true or false, although the procedure for deciding may be very lengthy and laborious.

Recently, examples have arisen in the study of quantum gravity[19] in which some quantity that is observable, in principle, is predicted to have a value given by a sum of terms that is uncomputable in Turing's sense (the list of terms in the sum could only be provided by a solution of a classic problem which is known to be uncomputable—the list of all compact four-manifolds). Again, while this may have a fundamental significance for our ability to predict, we cannot be sure that it imposes an unavoidable restriction. There may exist another way of calculating the observable in question using only conventional computable operations. Another interesting example of this sort arises in the biochemical realm, where it is known that Nature has found a way to fold complex proteins quickly. If it carries out this searching by the process that we would use to programme it computationally, then it would seem to be uncomputable in the same way that the 'travelling salesman problem' is uncomputable. However, Nature has found a way to carry out the computation in a fraction of a second. As yet we don't know how it is done. Again, we are led to appreciate the difference between limitations on Nature and limitations on the particular mathematical, computational, or statistical descriptors that we have chosen to use to describe its behaviour.

Summary

In this brief survey we have explored some of the ways in which the quest for a 'Theory of Everything' in the third millenium might find itself confronting impassable barriers. We have seen that there are limitations imposed by human intellectual capabilities, as well as by the scope of technology. There is no reason why the most fundamental aspects of the laws of Nature should be within the grasp of human minds, which evolved for quite different purposes, nor why those laws should have testable consequences at the moderate energies and temperatures that necessarily characterize life-supporting planetary environments. There are further barriers to the questions we may ask of the Universe, and the answers that it can provide us with. These are barriers imposed by the nature of knowledge itself, not by human fallibility or technical limitations. As we probe deeper into the intertwined logical structures that underwrite the nature of reality, we can expect to find more of these deep results that limit what can be known. Ultimately, we may even find that their totality characterizes the Universe more precisely than the catalogue of those things that we can know.

References

[1] This story seems to have arisen from a misinterpretation of a letter of resignation written by Commissioner Henry L. Ellsworth to his employers, the US Patent Office. The story is scotched by E. Jeffery in 'Nothing left to invent', *J. Patent Office*, 22, 479 (1940).

[2] J.D. Barrow, *Theories of everything*, Oxford University Press, Oxford (1991); J.D. Barrow, *The world within the world*, Oxford University Press, Oxford (1988).

[3] M. Green, J. Schwartz, and E. Witten, *Superstrings*, Cambridge University Press (1987); F.D. Peat, *Superstrings and the search for a theory of everything*, Contemporary Books, Chicago (1988); M. Green, 'Superstrings', *Scientific American*, September, p. 48, (1986); D. Bailin, 'Why superstrings?', *Contemp. Phys.*, **30**, 237 (1989); P.C.W. Davies and J.R. Brown (eds.), *Superstrings: a theory of everything?*, Cambridge University Press (1988).

[4] C. Froggatt and H.B. Nielsen, *Chaotic gauge theories*, World Scientific, Singapore; J.D. Barrow, and F.J. Tipler, *The anthropic cosmological principle*, Oxford University Press (1986).

[5] C. Isham, in *Physics, philosophy and theology*, eds R.J. Russell, W. Stoeger, and G.V. Coyne, Univ. Notre Dame Press, Indiana (1988); S.W. Hawking, *A brief history of time*, Bantam, New York (1988).

[6] M.B. Green and J. Schwartz, *Physics Letters B*, **149**, 117 (1984).

[7] J.D. Barrow, *The artful universe*, Oxford University Press, Oxford and New York (1995).

[8] J.D. Barrow, *Pi in the sky*, Oxford University Press, Oxford and New York (1992).

[9] L. Lederman, *The God particle*, Bantam, New York (1993).

[10] J.D. Barrow, 'Observational limits on the time-evolution of extra spatial dimensions', *Phys. Rev. D* **35**, 1805 (1987).

[11] J.D. Barrow, *The origin of the universe*, Orion, London (1994).

[12] A. Guth, 'The inflationary universe: a possible solution to the horizon and flatness problems', *Phys. Rev. D* **23**, 347 (1981); J.D. Barrow, 'The inflationary universe: modern developments', *Q. J. R,. Astron. Soc.* **29**, 101 (1988); A. Guth, and P. Steinhardt, 'The inflationary Universe', *Scientific American*, May, pp. 116–120, (1984); S.K. Blau and A. Guth, in *300 Years of Gravitation*, eds S.W. Harking and W. Israel, pp. 524–597, Cambridge University Press, Cambridge (1987).

[13] L. Krauss, *The fifth essence: the search for dark matter in the universe*, Basic Books, New York (1989).

[14] J.D. Barrow, 'Unprincipled cosmology', *Q. J. R. Astron. Soc.*, **34**, 117 (1993).

[15] J. Hartle and S.W. Hawking, 'The wave function of the universe', *Phys. Rev. D* **28**, 2960 (1983); J. Halliwell, 'Quantum cosmology and the creation of the universe', *Scientific American*, December, pp. 28–35 (1991).

[16] R. Penrose, *The emperor's new mind*, Oxford University Press, Oxford (1989).

[17] A. Vilenkin, 'Boundary conditions in quantum cosmology', *Phys. Rev. D* **33**, 3560 (1982).

[18] G. Chaitin, *Algorithmic information theory*, Cambridge University Press, Cambridge (1987); G. Chaitin, 'Randomness in arithmetic', *Scientific American*, July, pp. 80–85 (1980); J. Casti, *Searching for certainty*, Morrow, New York (1990).

[19] J. Hartle and R. Geroch, in *The quantum and the cosmos—J.A. Wheeler Festschrift*, ed. W. Zurek, Princeton University Press, Princeton (1988).

CHAPTER 11

Of the utmost gravity

What's the difference between a man who hits his head and a man who's head is hit? To a policeman quite a lot.

Reginald Hill, *Dalziel and Pascoe*

It all began one August day in 1684 when Dr Edmond Halley paid a visit to Isaac Newton's rooms in Cambridge; of that meeting Abraham de Moivre tells us that

> After they had been some time together, the Dr asked him what he thought the Curve would be that would be described by the Planets supposing the force of attraction towards the Sun to be reciprocal to the square of their distance from it. Sr Isaac replied immediately that it would be an Ellipsis, the Doctor struck with joy & amazement asked him how he knew it, Why saith he I have calculated it, whereupon Dr Halley asked him for his calculations without any further delay. Sr Isaac looked among his papers but could not find it, but he promised him to renew it, & then send it to him.

Three months later, true to his word, Newton sent a document to Halley that made good his remarkable claim. Encouraged further by Halley it took Newton but 14 months to transform and extend this manuscript into another entitled *Philosophiae Naturalis Principia Mathematica.* It was destined to become Book I of the *Principia.* Two further parts followed in March and April 1687 just two and a half years after Halley's first fortuitous visit.

The first edition of the monumental *Principia*, comprised of these three manuscripts, was published in London on 5 July 1687 under the imprimatur of Samuel Pepys, the President of the Royal Society (whose diary-keeping days were by then far behind him). A second edition appeared in 1713, and the third (and last) in 1726. Three years later the only full English translation ever made of the original Latin text was completed by Andrew Motte and published as a two-volume set. It is this chain of events that inspired the appearance of this

---

A review of *300 years of gravitation*, edited by S.W. Hawking and W. Israel, Cambridge University Press, 1987

tercentenary volume as a celebration of the most impressive and comprehensive work of science ever written. Today, no practising scientist reads the *Principia* out of necessity but it is the foundation stone upon which the entire edifice of mathematical physics and astronomy rests. Newton's laws of motion and gravitation, although superseded by Einstein's, provide an approximation to them that holds good in all the situations we encounter in everyday life. More important than the formulation of these rules by which Nature moves was the thoroughly rigorous and 'modern' approach that Newton introduced. The unity of nature was displayed by his ability to derive far-reaching and diverse results about the extraterrestrial physical world from a bundle of innocuous-sounding assumptions that stemmed from observations of mundane local objects such as apples and prisms. Despite its vast scope, to most scientists the *Principia* is synonymous with gravity and with it the first statement of a universal law of gravitation. This fact is reflected in the title and contents of Hawking and Israel's collection of 14 articles. With only one or two exceptions, the individual contributions are of the highest quality and expound their subjects in impressive style with careful attention to didactic completeness and recent historical development.

One of the strongest impressions to be gained from these articles is of the diversity of work in modern gravitation physics and the way in which gravitation is at last ceasing to be the aloof geometrical cousin of the rest of physics. In astrophysics it merges with the complexities of high-energy radiation processes to create spectacular physical phenomena. The search for gravitational waves exploits the latest technologies to provide astronomers with the promise of a new vista on the Universe. And in the inner space–time of elementary particles there are the first glimmerings of a convincing theory which might conjoin gravity with the other forces of Nature.

Who can doubt that Newton would be pleased but not surprised at these advances towards a unified description of the physical world and all its forces. For we find that 300 years ago, in the Preface to his stupendous book, he could write that, having determined 'the motions of the planets, the comets, the moon and the sea', he was unfortunately unable to determine the remaining structure of the world from the same propositions because

> I suspect that they may all depend upon certain forces by which the particles of bodies, by some causes hitherto unknown, are mutually impelled towards one another. These forces being unknown, philosophers have hitherto attempted the search of Nature in vain; but I hope the principles here laid down will afford some light either to this or some truer method of philosophy.

How right he was.

CHAPTER 12

Getting it together

The whole universe is one mathematical and harmonic expression, made up of finite representations of the infinite.

F. Kunz

Unity is a many splendoured thing. Whether you are advertising a unification church, a grand unified theory of physics, a united front, or merely galvanising a fractious political party for concerted campaigning, oneness is next to godliness. It is desirable. Diversity has its place, but things that are closely knit seem stronger and more reliable. All scientists regard the unity of the Universe as an unspoken presupposition that owes much to the great monotheistic faiths that underwrote Western science's faith in the rationality of Nature. A single Lawgiver means a single legislation: the decrees are what we call the laws of Nature. In subjects like astronomy and physics this has proved a stupendously successful working hypothesis. A small number of simple rules have enabled us to extend understanding of Nature from the inner space of the most elementary particles of matter to the outer space of the expanding universe of galaxies and quasars. We can understand the past and predict the future to an accuracy that the social sciences can only dream about. The unexpected ease with which we can compress so many of the superficial complexities of the Universe into simple patterns is at once an unexpected gift and the reason why we find mathematics to be so successful in telling us how the world works.

By contrast, if we look at the complexities of the living world around us—the life sciences, the brain sciences, the social sciences, economics, psychology—things look very different. Here, we find ourselves trying to predict the behaviours of things that are very different from atoms and molecules. Predicting the result of a chemical reaction does not alter that reaction in any way, but predicting the outcome of an election or the course of an economy certainly can.

The Universe appears to be governed by a small number of fundamental forces (four, it seems from what we have seen so far) whose form is dictated by symmetrical patterns. These few patterns determine the forms of the laws of Nature. Physicists believe they are on course to unify them into a single 'super-

---

A review of *Consilience: the unity of knowledge,* by E.O. Wilson, Alfred Knopf, New York, 1998

force'. Yet, the outcomes of those few laws—the staggering diversity of structures we see in the Universe—need not possess the same simple patterns as the laws themselves. That is the secret of the Universe: how it is possible for a small number of simple laws to produce such a never-ending abundance of asymmetrical complexity. We find some of those complexities when we start to explore the worlds of the human and social sciences. But often the fruits of our study of these worlds are rather unimpressive. Economists, psychologists, and social scientists make few reliable predictions. Even the most tumultuous human events, like the fall of the communist regimes in Eastern Europe or the great stock market crashes, came as complete surprises. Such is the difficulty of predicting the unpredictable.

Edward Wilson is a great entomologist who has never been afraid of stirring up a hornet's nest by applying what he has learned in his own field to other less developed areas of the behavioural sciences. In this book he spins his web around a word invented in 1840 by the British philosopher of science, William Whewell. 'Consilience' means lumping together or harmonising facts from different lines of investigation. Sciences like physics and chemistry are consilient: everything that we know about chemistry can be traced back to the physics of protons and electrons. Wilson believes that all human knowledge can be made consilient. He argues that there are two great barriers to achieving this single synthesis of human understanding. The first is that we don't know everything and this is actually an important stimulus to future inventiveness and discovery. He gives many elegant descriptions of what the near future of medical science may have in store for us. But the second barrier is more sinister. Wilson identifies a failure of many disciplines to appreciate the key insights made by the biological sciences. He laments economists' lack of any attempt to get to grips with human nature and natural selection. He bemoans the absence of any commitment by social scientists to the understanding of the workings of the human mind and the long evolutionary history that has made it what it is. In the gallery of the arts the same insights can help us understand why the human mind is inclined to like some images rather than others, to appreciate some patterns of sound more than others, and why we are attracted so powerfully by symmetry.

Taken in isolation these arguments are not new. There have been popular expositions of many of them in recent years, signalling the frustration of a new generation of scientists with the failure of traditional approaches to the study of the human mind and its cultural products. Wilson was one of the first biologists to attempt to use biological ideas to understand human (rather than animal) behaviour. Over the years he has wedded this vision to a deep concern for the future. The most impressive feature of this book is the way in which it culminates in a fervent appeal to all of us to consider anew the consequences of our voracious appetite for energy and other natural resources. We are creating a future for our children's children that will be lacking in food and water, polluted irreversibly, and denuded of innumerable species of flora and fauna. Wilson turns his fire on 'the myopia of most professional economists' as 'the single greatest intellectual obstacle to environmental realism'. He sees this as myopia as a consequence of their seduction by simple mathematical models of idealistic situations without a care for what we know of human psychology or biology. He cites, for example, the failure to include the full costs of human actions when evaluating the long-term viability of strategies which have powerful effects on the environment—the Brazilians could chop down the entire Amazonian rain forest and sell it as a lumber profit without any negative environmental entry appearing on their treasury's balance sheet.

This book should be read by all social scientists. It won't tell them how to solve the hard problems that they grapple with but it ought to inspire them to seek to make contact with what other, so far more successful, sciences have learnt. This is not to urge a blinkered reductionism upon the social sciences. Rather, we should appreciate that biology can inform psychology, and psychology can inform subjects, like economics, that include human action and bias. Unfortunately, in the past (as the reception to Wilson's famous book *Sociobiology* shows) many attempts to create cross-disciplinary fertilisation have stirred up hysterical responses by critics (often scientists in the same discipline as Wilson) who have particular political inclinations that they wish to see reflected in the human sciences. There is a tendency to associate all attempts at biological explanation in the study of human actions as odious genetic determinism. *Consilience* is not dissimilar to *Sociobiology* in its underlying philosophy and is in tune with the renamed son of sociobiology, so called 'evolutionary psychology'. But the writing is more memorable in its clarity, and the diversity of expertise that the author manages to marshal in support of his case is impressive. Although the first three chapters on the ancient history are a little uninspiring, the author becomes increasingly engaging and urgent as he seeks to persuade the reader of the power of a consilient viewpoint. I should stress that Wilson is not arguing for a single super-theory of everything, some equation which allows us to explain all art and human psychology in the way that Schrödinger's equation predicts the structure of all atoms. That would be naive. Rather, Wilson is arguing for a unification of the principles that are employed to understand the cultural behaviour of systems of living beings. Those principles may not be able to make simple predictions in every situation (after all, even meteorologists can't predict the weather with consistent accuracy) but they would unify our appreciation of what ingredients must play a role in a full understanding of the events we witness.

The worry I have about Wilson's embrace of what he calls 'empirical' rather than 'transcendental' explanations is that he fails to appreciate enough of the distinct approach of the fundamental physical sciences, where the precise mathematical forms of the laws of Nature determine things like the number and nature of the elementary particles, the global structure of the Universe, and the workings of gravity. The pre-eminence of mathematical simplicity and symmetry in understanding the forces of Nature is in effect an appeal to transcendental factors even though we then set about testing that appeal against all the evidence we can find. So far, no evidence has been found for the significant action of a principle like natural selection in these fundamental physical sciences. It is only in sciences of living things, where we study the complicated inter-relationships between different organisms, that the transcendental mode becomes weak and selection is a more powerful principle. However, it is still a subject of some debate whether natural selection alone can explain the origin of life or whether some physical 'laws' of complexity play a role. Either way, there is a real difference of approach between the fundamental physical and life sciences that is difficult to massage away by the mere invocation of a password like consilience.

It is because of this dichotomy that Wilson focuses upon the consilience of the studies of human cultural development. As an example of Wilson's approach, I would argue that there has long been a prejudice that the creative arts are an untrammelled expression of human creativity alone: unpredictable, arbitrary, and not subject to any sort of scientific understanding. But we know that there are ancient influences upon human sensitivities which have survival value. We have a predilection for certain types of landscape, certain statistical patterns of sound, and certain special symmetries. This cannot explain all our artistic leanings, but it can reveal why we are inclined to like certain things more than others. Nor is the process entirely one way, as Wilson might be interpreted as suggesting. Although the sciences of complexity can tell us interesting things about artistic patterns, it is also possible that the arts will provide wonderfully intricate examples of organised complexity for the sciences to understand.

The human mind is an extraordinary thing—the most complex entity we have encountered anywhere in this vast Universe. It has evolved to display a breathtaking precision in its conduct of operations too numerous to grasp. No single program can capture its power but that does not mean that it follows no rules and possesses no hardwired inheritance from millions of years of tuning that it has experienced by the evolutionary process, These rules and these inheritances are not all that the mind is: they are not all they we are, nor all we can be, but that is very different from saying they have no part to play in revealing to us how we think, what we are, and what we might be. It is here that Edward Wilson can help us: for he thinks more clearly than most about what we were, what we are, and what we might become if we fail to act responsibly.

Ingram Pinn

Part 4

Mathematics

Good mathematicians see analogies between theories;
great mathematicians see analogies between analogies.

Stefan Banach

Scientists have long known that there is safety in numbers. Mathematics offers a sure path to understanding Nature's workings. This feature of the world—the remarkable effectiveness of simple mathematics for its description—was one of the things that first attracted me to sciences like astronomy as a teenager. It seems quite extraordinary that simple patterns deduced from watching gases expanding in jars, or the swinging of a bob on the end of a piece of string, could tell us things about the structure of stars or the expansion of the Universe. Why was it that the next step the world took was the same as that predicted by squiggles on a piece of paper? Why does the Universe march to a mathematical tune?

If you want to know what mathematics is, a mathematicians is probably the last person to ask. Historians know what history is, sociologists know what sociology is, but most mathematicians neither know nor care what mathematics is: it is simply what mathematicians do. If you question your mathematical friends more closely, you will find a range of quite different views on offer: for some it is merely a game of logical patterns, like chess or chequers; for others it is the uncovering of a deep structure of reality, the nearest we get to thinking the unadulterated thoughts of God. In between these extremes there is plenty of room for compromise and variation.

If you want to probe further and discover why mathematics is so unreasonably effective in describing how the world works, then few mathematicians will offer a strong opinion. If one sees mathematics as the great catalogue of all possible patterns that there can be, then it becomes inevitable that the world 'is' mathematical. We could not exist in a patternless chaos. Any intelligible universe must therefore contain patterns, and where there are patterns there must also be mathematics. But this does not exorcise the real mystery. Why are such simple patterns so far-reaching in their explanatory power? It would be quite possible for the patterns of Nature to be described by mathematics and yet mathematics not be very useful to us in practice. This could be because natural patterns were of a sort that did not readily admit simple algorithms for carrying out computations, or because even the simplest natural patterns were to our understanding as quantum mechanics is to Schrödinger's cat.

The first chapter in this section, '*Why is the Universe mathematical?*' lays out the problem of 'what is mathematics?' and discuss some of the answers that have been suggested over the last century. The two newspaper articles '*It's all Platonic pi in the sky*' and '*Counter culture*' explore some of the human aspects of counting and mathematics: why we count as we do, and how our patterns of childhood play can help or hinder our aptitude for mathematics. We try to do away with the prejudices of the outsider that mathematicians are people who are 'good with figures', experts at mental arithmetic who spend their time adding up columns of figures—glorified accountants' clerks.

Finally, there is a short chapter entitled '*Rational vote doomed*' that appeared on general election day in Britain in 1997. In response to a request for an article about the election that was not about the election—of which, by then, everyone was thoroughly fed up with reading, and of which the result was a foregone conclusion—I wrote this piece about some peculiarities of democracy. It is possible for the sum of a number of rational individual choices to result in a paradoxical collective choice. The crux of the problem lies in relationships like 'preferring' or 'liking'. If Fred likes Jill and Jill likes Chris, this does not mean that Fred must like Chris. This is why human relationships are so unpredictable. By contrast, being 'taller than' necessarily possesses the required property, called transitivity, that if Fred is taller than Jill and Jill is taller than Chris, then Fred is taller than Chris. It turns out to be impossible to avoid this paradox without giving up some other desirable feature of a democratic voting system. Moreover, if our minds work by conducting some form of majority 'voting' as a result of many possible impulses to action being weighed up, then they must also deal with this little bit of collective irrationality.

CHAPTER 13

Why is the Universe mathematical?

Everything but mathematics must come to an end.

Paul Erdös

The most all-consuming category of thought that dominates our deepest images of the physical Universe is mathematics. Mathematics is the 'language' that allows us to talk most effectively, efficiently, and logically about the nature of things. But this mathematical language differs from other languages. It is not like English or Spanish. It is more like a computer language because it possesses a built-in logic. When we write a grammatically correct English sentence like 'All dogs have four legs and my table has four legs therefore my table is a dog' there is no guarantee that the sentence either be logically correct or have a correspondence with events in the world. Conversely, the grammatical incorrectness of a sentence like 'to boldly go where no man has gone before' does not render its realisation impossible. The rules of English grammar can be broken without sacrificing meaning. But break a rule of mathematics and disaster ensues. If any false mathematical statement is allowed it can be used to prove the truth of *any* mathematical statement. When Bertrand Russell once made this claim during a lecture he was challenged by a sceptical heckler to prove that the questioner was the Pope if twice 2 were 5. Russell at once replied that 'If twice 2 is 5, then 4 is 5, subtract 3; then $1 = 2$. But you and the Pope are 2; therefore you and the Pope are one'.

So mathematics is a language with a built-in logic. But what is so striking about this language is that it seems to describe how the world works not just sometimes, not just approximately, but invariably and with unfailing accuracy. All the fundamental sciences—physics, chemistry, and astronomy—are mathematical sciences. No phenomenon has ever been discovered in these subjects for which a mathematical description is not only possible but beautifully appropriate. 'A rose by any other name would smell as sweet' but not a rose by any other number. Yet one could still fail to be impressed. After the fact, we can force any hand into some glove and maybe we have chosen to pick the mathematical glove because it is the only one available. Yet, it is striking that physicists so often find some esoteric pattern, invented by mathematicians in the dim and distant past only for the sake of its elegance and curiosity value, to be precisely what is required to make sense of new observations of the world. In fact, confidence in mathematics has grown to such an extent that one expects (and finds) interesting mathematical structures to be deployed in Nature. Scientists look no further once they have found a mathematical explanation. Mathematics is no longer treated as a category of explanation, it has become the definition of explanation in the physical sciences.

There are many striking examples of the unexpected and unreasonable effectiveness of mathematics. In 1914 when Einstein was struggling to create a new description of gravity to supersede that of Newton he wished to endow the Universe of space and time with a curved geometry and write the laws of Nature in a form that would be found the same by any observers no matter what their state of motion. His old student friend, the mathematician Marcel Grossman, introduced him to a little-known branch of 19th-century mathematics that was tailor-made for his purposes. In modern times particle physicists have discovered that symmetry dictates the way elementary particles behave. Collections of related particles can behave in any way they choose as long as a particular abstract pattern is preserved. The laws of Nature are treated as the catalogue of habitual things that can occur in the world and yet preserve these patterns—in effect they define all those things that you can do to the world without changing its essence. To every such catalogue of changes we can ascribe an unchanging pattern although the pattern is often subtle and rather abstract. In the last century mathematicians set about investigating all the possible patterns that one could invent. These patterns are represented most basically by a branch of mathematics called 'group theory'. The catalogue of mathematical patterns that the early group theorists established has become the guiding force in the study of elementary particles. So successful were the simple patterns of the group theorists in describing the way the forces and particles of Nature behave that physicists have taken to exploring all the possible patterns in a systematic way to discover those that give the most interesting consequences if used in the construction of the Universe.

In this way the expectation that elegant mathematical ideas will be found in Nature allows detailed predictions to be made about the behaviour of elementary particles. Experiments at particle accelerators enable these aesthetic fancies to be tested against the facts. The theory of superstrings has become a focal point of interest and has also created a new phenomenon. For the first time particle physicists have run into concepts that require the invention of new mathematical structures and ideas to cope with them. For the first time 'off the shelf' mathematics is not enough to unravel all the patterns. Elsewhere, the beautiful subject of fractals has revealed itself to underwrite the whole spectrum of natural phenomena—from the clustering of galaxies to the crystalisation structure of snowflakes. Again, fractals were once no more than an idiosyncratic branch of mathematics investigated for its own sake. Fractal pictures offer striking images that resonate with our aesthetic sense of beauty. Presumably the reason again has roots in our evolutionary history. For we have discovered that many aspects of the Earth's natural scenery—mountain ranges, snowflakes, trees, and plants—that we find so visually appealing owe their structure, like fractals, to the process of self-similarity; that is, to the reproduction of a basic pattern on many different size scales.

Science seems to believe so deeply in the mathematical structure of Nature that it is an unquestioned article of faith that mathematics is both necessary and sufficient to describe everything from the inner space of elementary particles to the outer space of distant stars and galaxies—even the Universe itself. What are we to make of the ubiquity of mathematics in the constitution of the Universe? Is it evidence of a deep logic within the Universe: if so, where does that logic come from? Is it just a creation of our own minds or is God a mathematician?

We are confronted by a mystery. Why does the symbolic language of mathematics have everything to do with falling apples, splitting atoms, exploding stars, or fluctuating stock markets? Why does reality follow a mathematical lead? the sorts of answers we can offer to such puzzling questions depend crucially upon what we think mathematics actually is. There are four clear options—formalism, inventionism, constructivism, and realism.

At the beginning of this century mathematicians faced several bewildering problems which rocked their confidence. Logical paradoxes like that of the barber ('A barber shaves all those individuals who do not shave themselves. Who shaves the barber?'), highlighting the dilemma of the set of all sets that are not members of themselves, threatened to undermine the entire edifice. And who could forsee where the next paradox might surface? In the face of such dilemmas, David Hilbert, the foremost mathematician of the day, proposed that we cease worrying about the meaning of mathematics altogether. Instead, we should define mathematics to be no more and no less than the tapestry of formulae that can be created from any set of initial axioms by manipulating the symbols involved according to specified rules. This procedure, it was believed, could not create paradoxes. The vast embroidery of interwoven logical connections that results from the manipulation of all the possible starting axioms according to all the possible non-contradictory collections of rules then defines what mathematics 'is'. This is *formalism*.

Clearly, for Hilbert and his disciples the miraculous applicability of mathematics to Nature is something about which they neither care nor seek to offer any explanation. Mathematics does not have any meaning. The axioms and rules for the manipulation of symbols are not connected with observed reality in any necessary way. Formulae exist on pieces of paper but mathematical entities have no other claim to existence whatsoever. The formalist would no more offer an explanation for the mathematical character of physics than they would seek to explain why physical phenomena do not obey the rules of poker or blackjack.

Hilbert thought that this strategy would rid mathematics of all its problematical areas—by definition. Given any mathematical statement it would be possible, in principle, to determine whether it was a true or false conclusion from any particular set of starting assumptions by working through the network of logical connections. Hilbert and his disciples set to work confident that they could trammel up all known mathematics within this straightjacket. Unfortunately, and totally unexpectedly, the enterprise collapsed overnight. In 1931, Kurt Gödel, an unknown young mathematician at the University of Vienna, showed Hilbert's goal to be unattainable. Whatever set of consistent starting axioms one chooses, whatever set of consistent rules one adopts for manipulating the mathematical symbols involved, there must always exist some statement framed in the language of those symbols whose truth or falsity cannot be decided using those axioms and rules. Mathematical truth is something larger than axioms and rules. Try solving the problem by adding a new rule or a new axiom and you just create new undecidable statements. Check mate—Hilbert's programme cannot work. If you want to understand mathematics fully you have to go outside mathematics. If a 'religion' is defined to be a system of ideas that contains unprovable statements then Gödel has taught us that, not only is mathematics a religion, it is the only religion that can prove itself to be one.

Inventionism is the belief that mathematics is simply what mathematicians do. Mathematical entities, like sets or triangles, would not exist if there were no mathematicians. We invent mathematics: we do not discover it. The inventionist is not very impressed by the effectiveness of describing the world by mathematics. The reason that we find mathematics so useful is perhaps merely an indication of how little we know of the physical world. It is only the properties well suited to mathematical description that we have been able to uncover.

This view of mathematics is more commonly held by 'consumers' of mathematics—social scientists or economists for example—than by mathematicians themselves. It is closely associated with an idealist philosophical viewpoint. Although there may exist some ultimate reality, we cannot apprehend this except by filtering our observations and experience of it through certain mental categories which order it for our understanding. So, although we see the Universe to be mathematical, this does not mean that is really *is* mathematical any more than that the sky is pink because it looks that way when we wear rose-coloured spectacles.

If mathematics were entirely a human invention and used by scientists simply because it is useful and available then we might expect significant cultural differences within the subject. However, whereas there are discernible styles in the presentation of mathematics and in the type of mathematics investigated in different cultures, this is just a veneer. The independent discovery of the same mathematical theorems by different mathematicians from totally different economic, cultural, and political backgrounds at different times throughout history argues against such a simple view. Moreover, this unusual phenomenon of the independent multiple invention of the same mathematical truth sets creative mathematics apart from music or the arts. Pythagoras' theorem was independently discovered many times by different thinkers. It is inconceivable that Shakespeare's *Hamlet* or Beethoven's fifth symphony could be independently duplicated. This contrast argues that the foundation of mathematics lies outside the human mind and is not totally fashioned by our human way of thinking.

The most straightforward view of mathematics is to maintain that the world *is* mathematical in some deep sense. Mathematical concepts exist and they are discovered by mathematicians, not invented. Mathematics exists whether or not there are mathematicians. It is a universal language that could be used to communicate with beings on other planets who had developed independently of ourselves. For the realist, the number seven exists as an immaterial idea which we see realised in specific cases like seven dwarfs, seven brides, or seven brothers. This view is sometimes called *mathematical Platonism* because it assumes that there exists some other world of perfect mathematical forms which are the blueprints from which our imperfect experience is derived. Moreover, our mental processing of sense data is assumed to have a harmless effect upon the mathematical nature of reality. Realism of this sort seems tantamount to the view that God is a mathematician. And indeed, if the entire material Universe is described by mathematics (as modern cosmology assumes) then there must exist some immaterial logic that is larger than the material Universe.

The introduction of a Platonic interpretation of mathematics produces a striking parallel between mathematics and philosophical theology. The entire panoply of properties and attributes of God developed by early neo-Platonic religious philosophers can be taken over, almost word for word, to describe mathematics if we replace the word 'God' by 'mathematics'. The mathematics of the Platonist transcends the world and is viewed as existing both before and after the creation of the material world. When ancient philosophers tried to integrate concepts like the laws of Nature into a theological picture of the

Universe they succeeded without great difficulty. Moreover, they were able to incorporate suspensions of the laws of Nature, or miracles, in suitable ways. But the omnipotence of God rides awkwardly with mathematics. We can imagine a suspension or breach of a law of Nature (especially since we observe the outcomes of laws and not the laws themselves) but what about a breach of a law of logic or mathematics? Opinion was deeply divided over whether the omnipotence of God could run to creating a world where mathematical impossibilities exist. Spinoza believed that this freedom existed but ranged against him was the view that God did not have any such freedom of manoeuvre because it did not exist. This appears to make the deity subservient to the laws of mathematics and logic. Platonic mathematical reality was a rival to an omnipotent, omnipresent deity. One can pursue this argument further and discover analogues of dilemmas like the problem of evil, and even the issue of revelation versus reason in mathematical discovery, but this would take us far from our subject.

Realism regards the unreasonable effectiveness of mathematics in the description of Nature as crucial evidence in its support. Most scientists and mathematicians carry out their day-to-day work as if realism were true even though they might be loath to defend it too strongly at the weekend. But realism of this sort has a most extraordinary consequence. If we can conceive of any mathematical scheme for the evolution of the Universe in which observers like ourselves can exist (and clearly we can conceive of such a scenario) then this scenario exists in every sense of the world. Intelligent observers must exist.

Constructivism was another response to the climate of uncertainty about logical paradoxes that spawned formalism in the early years of this century. Its starting point, according to Kronecker, one of its creators, was the recognition that 'God made the integers, all else is the work of man'. What he meant by this was that we should accept only the simplest possible mathematical notions—that of the whole numbers 1,2,3,4, . . . and counting—as a starting point and then derive everything else from these intuitively obvious notions step-by-step. By taking this conservative stance the constructivists wished to avoid encountering or manipulating entities like infinity sets about which we could have no concrete experience and which have counter-intuitive properties (infinity minus infinity = infinity, for instance). Mathematics now consists of the collection of statements that can be constructed in a finite number of deductive steps from the starting point of whole numbers. The 'meaning' of a mathematical formula is simply the finite chain of computations that have been used to construct it. This view may sound harmless enough but it has dire consequences. It creates a new category of mathematical statement. Its status can now be true or false or undecided. A statement whose truth cannot be decided in a finite number of constructive steps is given this latter, limbo, status. The most important consequence of this policy is that a statement about infinite sets is no longer either true or false.

Pre-constructivist mathematicians had developed all manner of ways of proving formulae to be true that do not correspond to a finite number of constructive steps. One famous method beloved of the ancient Greeks is the *reductio ad absurdum*. To show something to be false we assume it to be true and from that assumption deduce something contradictory (like $2 = 1$). From this we conclude our original assumption to have been false. This argument is based upon the presumption that a statement that is not true is false. This becomes an invalid move under the constructivists' rules. The whole body of mathematical theorems that prove that something exists, but do not construct an example of it explicitly, are outlawed.

This philosophy of mathematics would have interesting but largely unexplored consequences if adopted in physics because many important physical theories, like Einstein's general relativity or Niels Bohr's quantum mechanics, make important use of non-constructive reasoning in deducing properties of the Universe. To most mathematicians such a strategy seems rather depressing, tantamount to fighting with one arm tied behind your back.

In general, all physicists use non-constructive mathematical arguments without a second thought. The only area of physics where the restriction to constructive methods and its accompanying three-valued logic has been investigated closely is in the quantum measurement problem where it has been proposed as a way of accommodating the issues raised by the Einstein–Podolsky–Rosen paradox. Yet, if a constructivists view is, at root, the correct one, it radically affects our attempts to deduce a 'Theory of Everything'. How such a view might be forced upon us by adopting a computational paradigm for the Universe we shall discuss shortly.

There have always been passionate advocates of the constructivist interpretation of mathematics. One particularly dogmatic supporter was the leading Dutch mathematician Luitzen Brouwer who, whilst an editor of the German journal *Mathematische Annalen* (the foremost mathematical journal of the day) declared war on non-constructivist mathematicians by rejecting any papers submitted for publication that used non-constructive methods, infinities, or the *reductio ad absurdum*. This created a considerable rumpus amongst mathematicians. The other members of the editorial board resolved the crisis by resigning *en bloc* and then recreating a new editorial board—excluding Brouwer. The Dutch government viewed this as an insult to their distinguished countryman and responded by creating a rival journal with Brouwer as editor.

Now what has constructivism to say about the mathematical character of Nature? We can see that it inherits something of what remained of Hilbert's formalist programme following its devastation by Gödel's discovery. We learn that there must always be some statements whose truth we can neither prove nor disprove, but what about all those statements whose truth we can decide by the traditional methods of mathematics. How many of them could the constructivists prove? Can we build, at least in principle, a computer that reads input, displays the current state of the machine, and possesses a processor for determining a new state from its present one, and use it to decide whether a given statement is true or false after a finite time? Is there a specification for a machine that will decide for us whether all the decidable statements of mathematics are either true or false? Contrary to the expectations of many mathematicians the answer was again *no*. Alan Turing in Cambridge, and Emil Post and Alonzo Church in Princeton showed that there were statements whose truth would require an infinite time to decide. They are, in effect, infinitely deeper than the logic of step-by-step computation. The idealised computer we have outlined is called a *Turing Machine*; it is the essence of every computer. No real computer was believed to possess greater problem-solving ability.

The mathematical operations that a Turing machine cannot perform in finite time are called *non-computable*. Many examples are known and could have some interesting physical consequences. We do not know whether Nature incorporates non-computable things into its fabric. If, for example, the action of the human mind or the phenomenon of human consciousness involve non-computable operations, then the quest for artificial intelligence cannot succeed in producing computer hardware able to mimic the complexity of human consciousness completely. Whether such a restriction would be of any practical interest depends of course upon how crucial the non-computable aspects are to brain function. At present it seems unlikely that they are.

If we return to the puzzle of the applicability of mathematics to Nature we can recast it into an interesting statement about computability. If an operation is computable it means that we can fabricate a device from matter whose behaviour mimics that operation. Typical devices might be swinging pendulums or electrical circuits. By the same token, physical devices like these can be well described by computable mathematical operations. The fact that Nature is well described by mathematics is equivalent to the fact that the simplest mathematical operations, like addition and multiplication, along with the more complicated operations used so effectively in science, are computable functions. If they were not then they could not be equivalent to any natural process and we would not be terribly impressed by the practical usefulness of mathematics.

It is a fascinating question to ask whether or not the laws of Nature contain non-computable elements. Already the quest to create a quantum theory of the whole Universe has created this possibility. There exist potentially observable attributes of the Universe defined by infinite sums of terms that possess no computable listing. They cannot be listed by any systematic calculation that just applies the same principles over and over again. Each entry requires qualitatively different and novel principles to be used for its itemisation.

There are two great streams of thought in contemporary science that, after running in parallel for so long, have begun to discover tantalising channels that point to their future convergence. The circumstances of this coming together will determine which of them will ever after be seen as a mere tributary of the other. On the one side is the physicists' belief in the 'laws of Nature', guilded with symmetry, as the most fundamental bedrock of logic in the Universe. These symmetries are wedded to the picture of space and time as indivisable continua. Set against this is some picture of abstract computation, rather than symmetry, as the most fundamental of all notions. This image of reality portrays the logic at its bedrock as governing something discrete rather than continuous. The great unsolved puzzle for the future is to decide which is more fundamental—symmetry or computation. Is the Universe a cosmic kaleidoscope or a cosmic computer: a pattern or a program? Or neither? The choice requires us to know whether the laws of physics constrain the ultimate capability of abstract computation. Do they limit its speed and scope? Or do the rules governing the process of computation control what laws of Nature are possible?

Before discussing what little we can say about this choice it is best to be on ones' guard concerning the choice itself. Throughout the history of human thought there have been dominant paradigms for the Universe. These mental images often tell us little about the Universe but much about the society that was engaged in its study. For those early Greeks, who had developed a teleological perspective on the world as a result of the first systematic studies of living things, the world was a great organism. To others, who held geometry to be revered above all other categories of thought, the Universe was a geometrical harmony of perfect shapes. Later, in the era when the first clockwork and pendulum mechanisms were made, the image of the post-Newtonian Universe as a mechanism held sway and launched a thousands ships of apologetics in search of the cosmic clockmaker. For the Victorians of the industrial revolution the prevailing paradigm was that of the heat engine and the physical, and philosophical questions it raised concerning the laws of thermodynamics and the ultimate fate of the Universe bear the stamp of that age of machines. So, today perhaps the image of the Universe as a computer is just the latest predictable extension of this habit of thought. Tomorrow there may be a new paradigm. What will it be? Is there some deep and simple concept that stands behind logic in the same manner that logic stands behind mathematics and computation?

At first, the motions of symmetry and computation seem far removed and the choice between them appears a stark one. But symmetries dictate the possible changes that can occur and the 'laws' that result might be viewed as a form of software which runs on some hardware—the material 'hardware' of our physical Universe. Such a view implicitly subscribes to one of the particular views of the relationship between the laws of Nature and the physical Universe that we introduced in Part 2. It regards the two as disjoint, independent conceptions. Thus, one could envisage the software being run upon different hardware. This view, then, seems to lead us into potential conflict with a belief in some unique 'Theory of Everything' that unites the conditions for the existence of elementary particles to the laws that govern them.

The success of the continuum view of the physicists in explaining the physical world appears at first sight to argue against the discrete computational perspective. But logicians have waged a war of attrition against the notion of the number continuum over the last 50 years. Logicians like Quine claim that 'Just as the introduction of the irrational numbers . . . is a convenient myth [which] simplifies the laws of arithmetic . . . so physical objects, are postulated entities which round out and simplify our account of the flux of existence . . . The conceptual scheme of physical objects is a convenient myth, simpler than the literal truth and yet containing that literal truth as a scattered part.'

As yet we have not found the right question to ask of the Universe whose answer will tell us whether computation is more primitive that symmetry: whether, in John Wheeler's words we can get

IT from BIT

Probably, this hope cannot be completely fulfilled. In order for it to be found that computation is the most basic aspect of reality we would require that the Universe only do computable things. The scope of the Universe's mathematical manifestations would be constrained to lie within the remit of the constructivists. This is the penalty of giving up the continuum and appealing to computable aspects of the world as the basis for explaining the whole. Yet we have uncovered many non-computable mathematical operations and physicists have found that many lurk within that piece of mathematics that is currently required for our understanding of the physical world.

The answers to these difficulties, if they can be found, surely lie in an enlarged concept of what we mean by a computation. Traditionally, computer scientists have defined the ultimate capability of any computer—whether real or imaginary—to be that of Turing's idealised machine. Indeed, the capability of such a machine defines what we mean by the accolade 'computable'. Yet, in recent years, it has become clear that one can fabricate computers that are intrinsically quantum mechanical in nature and so exploit the quantum uncertainties of the world to perform operations that might be beyond the capability of Turing's idealised machine. Since the world is ultimately a quantum system, any attempt to explain its inner workings in terms of the computational paradigm must be founded upon a firm understanding of what quantum computation actually is and what it can achieve that a conventional Turing machine cannot. In many ways the computational paradigm has an affinity for the quantum picture of the world. Both are discrete; both possess dual aspects like evolution and measurement (compute and read). But greater claims could be made for the relationship between the quantum picture and the symmetries of Nature. Half a century of detailed study by physicists has wedded the two into an indissoluble union. What might be the status of the computational paradigm after a similar investment of thought and energy?

The computer invites us to abstract its essence. Stripped of its accoutrements of hardware and task-specific software, it is a Turing machine—a processor of information. A mapping of one finite string of integers into another. We might take this image with a Kuhnian pinch of salt, as the latest in a never-ending sequence of fashions of explanation to be discarded when the next technological revolution occurs. Aware of this possibility, let us, none the less, assume computation to have non-ephemeral significance. We can ask ourselves whether it is more basic to view the evolution and structure of the Universe as a computation or as the consequences of laws of Nature. Or, merging the two concepts, whether we should treat the laws of Nature as though they are a form of software that happens to be running upon the material content of our Universe. Whereas the picture of laws of Nature as symmetries and invariances so beloved by the physicist blends naturally with the Platonic view of mathematical reality, the computational picture seems to point more naturally to the more limited constructivist view.

The most fruitful outcome from a computational image of Nature is that it reveals the bare bones of why Nature is intelligible to us, why science is possible, and why mathematics is so effective in the description of the physical world.

If we are presented with a sequence of numbers or symbols then it may be possible to replace the list by an abbreviated statement that has identical information content. Thus, the infinite sequence of numbers 2,4,6,8,10, . . . could be replaced by the formula for generating even numbers. We say that our sequence is *algorithmically compressible*. A random sequence is characterised by the fact that there is no formula shorter than itself that encapsulates it. (It is an interesting consequence of Gödel's undecidability theorem that one cannot prove a sequence to be random in this sense.) Truly random sequences cannot be compressed into simplifying formulae. They are algorithmically incompressible: defined by nothing less than their own listing. Science exists because the natural world appears to be algorithmically compressible. The mathematical formulae that we call laws of Nature are economical compressions of huge sequences of data about how the states of the world change. This is what it means for the world to be intelligible. We can conceive of a world in which all phenomena are chaotically random (just as some of them are seen to be). Its properties could only be described by listing innumerable time sequences of observed phenomena. Science would then be more like train-spotting. Observed phenomena would have that uniqueness which we find in the world of creative art. If the Universe is a unique and necessary entity then we might not be surprised to find the Universe as a whole to be an algorithmically incompressible entity: ultimately irreducible to any abbreviated formula, defined most simply by nothing less than its complete unfolding sequence of events. The search for a 'Theory of Everything' is the ultimate expression of our faith in the algorithmic compressibility of Nature.

We see science as the search for algorithmic compressions of the world of experience and the search for a single all-encompassing 'Theory of Everything' as the ultimate expression of some scientists' deeply held faith that the essential structure of the Universe as a whole can be algorithmically compressed. But we recognise that the human mind plays a non-trivial role in this evaluation. Inextricably linked to the apparent algorithmic compressibility of the world is the ability of the human mind to carry out compressions. Our minds have evolved out of the elements of the physical world and have been honed, at least partially, towards their present state by the perpetual process of natural selection. Their effectiveness as sensors of the environment and their survival value are obviously related to their abilities as algorithmic compressors. The more efficiently they can store and codify an organism's experience of the natural world, the more effectively can that organism counter the dangers that an otherwise unpredictable environment presents. In our most recent phase of history as *Homo sapiens*, this ability has attained new levels of sophistication. We are able to think about thinking itself. Instead of merely learning from experience as part of the evolutionary process, we have sufficient mental capacity to be able to simulate or imagine the likely results of our actions. In this mode our minds are generating simulations of past experience embedded into new situations. But to do this effectively requires the brain to be rather finely balanced. It is obvious that mental capacity must be above some threshold in order to achieve effective algorithmic compression. Our senses have to be sensitive enough to gather a significant amount of information from the environment. But it is understandable why we have not become too good at this. If our senses were so heightened that we gathered every piece of information possible about the things that we saw or heard—all the minutae of atomic arrangements—then our minds would be overloaded with information. Processing would be slower, reaction times longer, and all sorts of additional circuitry would be required to sift information into pictures of different levels of intensity and depth.

The fact that our minds are not too ambitious in their information gathering and processing abilities means the brain would effect an algorithmic compression upon the Universe whether or not it were intrinsically so compressible. In practice, the brain does this by truncation. Our unaided senses are only able to take in a certain quantity of information about the world down to some level of resolution and sensitivity. Even when we enlist the aid of artificial sensors like telescopes and microscopes to enlarge the range of our faculties there are still fundamental limits to that extension. Often this truncation process becomes rather formalised as a branch of applied science in itself. A good example is statistics. When we study a large or very complicated phenomenon we might try to compress the information available by sampling it in some selective way. Thus, pollsters trying to predict public opinion before a general election should ask every individual in the country who they will vote for. In practice, they ask a representative subset of the population and almost invariably produce a startlingly good prediction of the results of the full election.

Mathematics is useful in the description of the physical world because the world is algorithmically compressible. It is the language of sequence abbreviation. The human mind allows us to make contact with that world because the brain possesses the ability to compress complex sequences of sense data into abbreviated form. These abbreviations permit the existence of thought and memory. The natural limits that Nature imposes upon our sensory organs prevent us from being overloaded with information about the world. These limits act as a safety valve for the mind. Yet we still owe everything to the brain's remarkable ability to exploit the algorithmic compressibility of the world. And most remarkable of all, the brain is an evolved complex state of the very world whose complexity it seeks to compress, albeit one that has yet to fathom its own complexity.

Bibliography

J.D. Barrow, *The world within the world*, Oxford University Press, Oxford (1988).

J.D. Barrow, 'The mathematical universe', in *The World and I*, May, p. 306 (1989).

J.D. Barrow, *Theories of everything: the quest of ultimate explanation*, Oxford University Press, Oxford (1991).

J.D. Barrow and F.J. Tipler, *The anthropic cosmological principle*, Oxford University Press, Oxford (1986).

G. Chaitin, *Algorithmic complexity*, Cambridge University Press, Cambridge (1988).

D. Deutsch, 'Quantum theory, the Church–Turing principle, and the universal quantum computer', *Proc. Roy. Soc. A* **400**, 97–117 (1985).

B. Mandelbrot, *The fractal geometry of nature*, W.H. Freeman, San Francisco (1982).

R. Penrose, *The emperor's new mind*, Oxford University Press, Oxford (1989).

A. Turing, 'On computable numbers, with an application to the Entscheidungproblem', *Proc. London Math. Soc.* 42, 230, 546 (1936).

E. Wigner, 'The unreasonable effectiveness of mathematics in the natural sciences', *Commun. Pure Appl. Math.* 13, 1 (1960).

CHAPTER 14

It's all Platonic pi in the sky

Most mathematicians feel like a hunter in the jungle, where the theorems sit in the trees or fly around, where the definitions are like convenient ladders awaiting to be used in order to trap the theorems and corollaries which are there, even if nobody finds them.

Emilio Roxin

The growth of science has been wedded to a faith in the effectiveness of mathematics as a description of the Universe. Its utility is a secret whose recognition opened the door to our understanding and a manipulation of Nature.

We do not fully know why mathematics works. Some believe the Universe 'is' mathematical because mathematical entities exist in some Platonic 'pi in the sky' sense: when Pythagoras proved his famous theorem it was an act of discovery rather than of creativity. Physicists have always been struck by the unreasonable utility of mathematics as a description of the physical world. Time and again they have found the esoteric pieces of so-called 'pure' mathematics, written long ago by mathematicians interested only in the internal harmony and elegance of the subject, turn out to be precisely the language in which some newly discovered facet of Nature is most naturally expressed. Their amazement is heightened by the fact that mathematics in Nature involves the subtlest structures that are the furthest removed from everyday experience. Moreover, lest one think that current mathematical explanations amount to retracing our own footprints in the sands of time, it works most powerfully in the astronomical and elementary particle realms, an understanding of which appears to have offered little or no past advantage in the evolutionary hurly-burly. What we learn is that Nature is expressed in the language of mathematics —a language that differs in crucial respects from a language like English. Make a grammatical mistake and we can still communicate. But break the rules of logic and all is lost.

Mathematics is like a computer language because it has a built-in logic whose neglect empties it of meaningful content. It may be helpful in the early stages to teach English in a way that overlooks spelling errors and grammatical infelicities. But to apply such an attitude to logical accuracy when teaching mathematics would spell disaster.

What is it that mathematicians study? At root it is nothing less than all possible patterns; all the conceivable relationships between things. Some of these things are familiar objects; others are only mental pictures. The scope of pattern recognition is all-encompassing. Consider the multitude of jobs that rely on seeing and exploiting patterns: in sales trends, accident statistics, learning abilities, voting trends, money markets, electrical faults or human needs. Mathematicians study patterns devoid of specific content: they are interested in generalities that can then be applied to particular situations.

The ability to appreciate mathematical concepts may crystallise in our minds during the pre-school years. Psychologists have emphasised the formative aspects of infant play that establish intuition about notions like sameness and difference, addition and subtraction. Playing with bricks teaches important lessons about geometry, logic and pattern. Years ago it would have been common to find a boy's bedroom filled with toys with components that fit together according to definite geometrical rules. The young girl's bedroom, on the other hand, would have abounded with fluffy toys. Not a lesson to be learned about the structure of space or the rules of logical combination. Nowadays all young children seem fixated by computer games. It is easy to think that all exposure to computers produces intuition and confidence about them. But beware; you would hardly argue that years spent in snooker halls encourages intuition about Newtonian mechanics.

The mathematical language is remarkable but its study is often isolated. Educators miss many opportunities to alleviate the undoubted difficulty of learning mathematical concepts by mingling them with the history and anthropology that surrounds them. The number symbols we use were first introduced in ancient India. They represent the greatest intellectual innovation in history. They are far more universal than the letters of the Phoenician alphabet in which this sentence is written. They are a masterpiece of economical information storage. With just 10 symbols, any quantity can be represented. The secret is the 'place value' trick where the relative position of the symbols carries information. Thus 11 means 10 plus one. In a system without this nuance, like Roman numerals, the relative positions of the symbols do not carry information. The result is a cumbersome system in which even the multiplication of the two Roman numerals, like CVII and LXIII, is a major enterprise because the notation does not think for you in the way that it does when we multiply 107 by 63. The lesson from this digression is simply that the origins of counting, of our number symbols, the beginnings of geometry in architecture and surveying, the logical games like cat's cradle, are important. They bring out the universality of mathematical concepts. They help show why we do mathematics in the way that we do, why the words and symbols we use for numbers are as they are and how the activities of modern mathematicians differ from mere counting.

Mathematics can even be a matter of life or death. During the Russian revolution, the mathematical physicist Igor Tamm was seized by anti-communist vigilantes at a village near Odessa where he had gone to barter for food. They suspected he was an anti-Ukranian communist agitator and dragged him off to their leader. Asked what he did for a living, he said he was a mathematician. The sceptical gang leader began to finger the bullets and grenades slung around his neck. 'All right', he said, 'calculate the error when the Taylor series approximation of a function is truncated after *n* terms. Do this and you will go free. Fail and you will be shot.' Tamm slowly calculated the answer in the dust with his quivering finger. When he had finished, the bandit cast his eye over the answer and waved him on his way. Tamm won the 1958 Nobel prize for physics and he never did discover the identity of the unusual bandit leader. But he found a sure way to concentrate his students' minds on the practical importance of mathematics.

CHAPTER 15

Counter culture

I recommend you to question all your beliefs, except that two and two make four.

Voltaire

Say, 'mathematician' to the man on the Clapham omnibus and he conjures up images of an assiduous book-keeper or an old-fashioned green-grocer, someone good at figures, at adding up columns of numbers, someone on whom you can count. To the outsider, mathematics is only about numbers—and, once, this might have been the case. Many ancient cultures displayed a deep affinity for numbers in themselves. Each had a meaning and a cosmic significance: lucky numbers, unlucky numbers, numbers which are symbolic, masculine or feminine, the strange religious taboos of the census found in disparate cultures, don't count your chickens before they hatch, your number may be up. How strange that seemingly irrational beliefs should be so widespread.

Modern mathematics sees numbers quite differently. The path from numerology to numeracy has been a gradual awareness that it is in the *relationships* between numbers, not the numbers themselves, that deep significance lies. Counting, geometry, algebra and all the other parts of mathematics are simply explorations of different types of relationships and patterns. Mathematics is the study of all possible patterns. Some of those patterns have concrete expressions in the world around us. We see spirals, triangles, circles and squares. Others are abstract extensions of these worldly examples, and more still seem to live only in the fertile minds of their conceivers. Viewed like this, we see why there has to be something akin to mathematics in the Universe. We and any other sentient being are, at root, examples of organised complexity. We are complex, stable patterns in the fabric of the Universe. If there is to be life in any shape or form there must be a departure from randomness and total irrationality. Where there is life there is pattern and where there is pattern there is mathematics. Once that germ of rationality and order exists in a potential cosmic chaos, then so does mathematics. There could not be a non-mathematical Universe that contains living observers.

There could exist only a modicum of order at the heart of things. That part of the whole that interfaces with our own parochial evolutionary history might be nothing but a drop in a sea of rationality. Alternatively, the order behind the world and the language of mathematics needed to describe it might have been of a deep and uncomputable complexity. The patterns of Nature might have been unfathomable by any conscious subset of those patterns. In such a world, mathematics would possess no utility for predicting the future of the heavens and the Earth. But perhaps again we could not be in such a convoluted creation. To evolve successfully in a complex patterned natural environment our pre-conscious minds had to create a model of that environment: an embodied 'theory' of it whose disproof would lead to extinction.

For living complexity to evolve successfully it must be able to store information about that environment and carry out, at first, simple and later, complicated computations. The success of this process relies upon a bedrock of reliable pattern that can be approximated without fatal consequences. Seen in this light, the existence of a certain level of pattern and discernible order in the natural world is neither unexpected nor mysterious. At least, no more, and no less, than our own existence. But when we look elsewhere things become distinctly odd. When we probe the inner space of the most elementary particles of matter or the outer space of galaxies and stars we find that mathematics is wonderfully effective in its descriptions of what is to be found there. It is appropriate. It is unrivalled. And the mathematics upon which one can count is not just the mundane and the familiar: it is the deepest and most abstract of the creations that the pure mathematicians—those poets of its language—have fashioned from the slightest promptings of reality. Time and again it has been found that these other-worldly studies have provided the physicist with precisely the logic and pattern required to describe and predict Nature's most deeply laid and glittering mechanisms. So successful has the mathematical modelling of Nature been in the realms of fundamental physics that the ultimate theories of physics are sought in no other place than in the catalogue of beautiful mathematical structures. And their beauty lies where much more of that virtue is to be found: in the unification of superficial diversity into a single simplicity, in the replacement of the many and the possible by the one and only.

Very occasionally and, our experience teaches us, only when the deepest truths behind the Universe's workings are at hand, the physicist exhausts the supply of off-the-shelf mathematics and must seek new mathematical patterns to further the understanding of Nature. Such a vacuum once led Newton to the calculus. More recently the desire of particle physicists to understand the first consistent 'Theories of Everything' has revealed new areas of mathematics illuminated by the brilliance of the logical patterns of the interwoven forces of Nature. As the Universe expanded and cooled, those symmetries were hidden and disguised. Simplicity was overtaken by complexity and the mathematical beauty disguised. So great is physicists' faith in the ubiquity and utility of mathematics that they look to the workings of Nature to learn new mathematics. Laws of Nature are simply the invariant mathematical patterns instantiated in the Universe. But still we marvel at the weird appropriateness of the mathematics which our predecessors created with no intention of describing the way the world works.

The unexpected effectiveness of mathematics in these realms is counter to much popular opinion. 'Pi' may be in the sky but many still see mathematics as little more than what mathematicians do: an activity reserved for those sartorially challenged Open University chaps on the telly. But the further we go from the realms of direct human experience, and from those aspects of reality whose accurate preconscious representation or conscious comprehension plays an important role in our evolutionary history, the better we find mathematics to work. In Victorian times there was a certain robust commonsense philosophy of science which regarded scientific inquiry as self-evidently being about something real. A science library would contain weighty tomes with severe titles like 'The Theory Of Sound'. Today we are more likely to find 'Mathematical Modelling Of Sonic Phenomena' on the shelves, filling a much-needed gap in the literature. Implicit in the latter is an emphasis upon mathematics as a way of learning something—and far from everything—about a thing called 'sound'. The emphasis is upon 'modelling'—that is, producing a partial, simplified version of the real thing—rather than on the unfolding of 'the' complete theory of something real.

There is a good reason for this unsoundness. The things that mathematical modellers study are complicated. They include things, such as societies and economies, that are even more complicated than sound because they respond to being studied. Weather patterns do not change because you forecast them. Voting patterns do. So do economic trends. Modelling these self-interactive situations is a compromise with unmanageable complexity created by the simultaneous influence of many different interacting factors, none of which may be very subtle when acting in isolation. If simple symmetries and patterns can be found which approximate what goes on in reality, then the idealised picture of reality they create may often be very useful in studying the true state of affairs.

This is not always a useful strategy because the smallest error of description of what is being studied may quickly amplify into a huge disparity with its true behaviour. Such sensitive phenomena—and the weather is one of them—are called chaotic. Such phenomena by no means defy mathematical description. They merely lead us to describe them by means of new types of mathematics and to be aware of the limitations upon predictability in practice even when it exists in principle.

Viewed like this, the world might seem to be nothing but a display of mindless mathematical order. Indeed, you can be a dehumanist if you choose. A work of Rembrandt or Miro, of Mozart or Cage, can always be viewed only as an elaborated form of mathematical pattern if you insist. But while it is possible to analyse musical works in terms of variations in air pressure or paintings as frequency variations in light, it is not terribly useful. These things are somehow only diluted by their superficial encoding in mathematical patterns. Indeed, it is the very fact of this vast subjective realm of complexity, defying quantification for any useful purpose, that all the more impresses on us the miracle of a Universe that elsewhere is so totally, deeply and unerringly mathematical.

CHAPTER 16

Rational vote doomed

A completely unfree society—that is, one proceeding in everything by strict rules of "conformity"—will, in its behaviour, be either inconsistent or incomplete, unable to solve certain problems, perhaps of vital importance.

Kurt Gödel

Nothing looks simpler than voting. But then nothing always has been a dodgy idea. It has long been known that there is something distinctly worrying about democracy—and I don't mean that it can fall into the wrong hands. From a distance it looks like the totalling up of lots of individual choices to arrive at the will of the people. What could be fairer and simpler? But imagine a small world of three voters, Alex, Bill, and Chris, who can express their preferences for three policies regarding social priorities for expenditure on defence, education, and welfare. Alex's order of preference is defence, education, and welfare; Bill's is education, welfare, and defence; while Chris's is welfare, defence, and education. Nothing odd so far. The tellers add up the votes to see which order of preference is the electorate's collective will. Defence has beaten education by two votes to one, education has beaten welfare by two votes to one. The trend looks pretty clear. But hang on, welfare has beaten defence by two votes to one! Are we all going mad? Maybe there is some truth in the rumour that in an autocracy, one person has his way; in an aristocracy a few people have their way; but in a democracy no one has his way.

This paradoxical outcome is extremely worrying. Discovered by the French social scientist, the Marquis de Condorcet, in 1785, it shows that rational individual choices can lead to unreliable collective decisions. As we try to pass from individual choices to some form of collective choice, a paradox arises. Social choices are quite different beasts to individual choices, despite being composed of nothing more than individual choices. As a result, collective social choices sometimes exhibit an arbitrariness that does not reflect the way that personal decisions are made. Personal decisions arise from individual inclinations and tastes, but collective social choices do not: society does not have an inclination or a taste of its own.

Alex, Bill, and Chris seem to be going round in circles because preferences are relationships that do not possess the simple property of 'transitivity'. Some relationships do. For example, being 'older than' does: if Alex is older than Bill, and Bill is older than Chris, then Alex must be older than Chris. But 'liking' someone does not: if Alex likes Bill, and Bill likes Chris, then this does not mean that Alex must like Chris—a truth that any family reunion will confirm.

One response to these paradoxes might be to hope and pray that intransitive preference patterns, like those displayed by Alex, Bill, and Chris, will arise very rarely. So, how likely is it that they do? The three voters each have six possible ways of ordering their preferences, and so there are a total of $6 \times 6 \times 6 = 216$ possible voting patterns. Of these, twelve lead to paradoxical collective choices. So, the probability of a paradox is roughly six per cent; small, but not negligible. Increase the population of the electorate from three to as many as you like, and it rises to nearly nine per cent. But increase the number of alternatives that three or more voters have to choose from, and the probability of paradox quickly soars to one hundred per cent—complete certainty.

These voting paradoxes were once regarded as curiosities that could always be avoided in real life, rather like logical paradoxes of the 'this statement is false' variety that used to worry logicians before they distinguished clearly between statements *of* logic and statements *about* logic. But things took an unexpected turn in 1950 when the American economist Kenneth Arrow analysed the problem of democratic choice in a completely general and transparent fashion. The result, which helped earn Arrow the Nobel prize in economics in 1972, is known as the Arrow Impossibility Theorem by the pessimists, and the Arrow Possibility Theorem by the optimists.

Arrow showed that any election obeying a few simple general requirements that we would expect of any democratic system, will create this paradox of collective choice. If there is complete individual freedom of choice, if the result of the election cannot arise independently of voters' choices (so it is not imposed on society by some outside influence, like a religious system), and if there is no dictator whose single vote determines the overall social choice, then it is impossible to avoid a voting paradox arising. Only by dropping one or more of the democratic requirements can paradox be side-stepped.

Today, voting is not confined to electing parish councils, black balling candidates for the Athenaeum, judging ice-skating competitions, or even electing governments. Advanced technological systems, like space launches, are often under the control of a number of computers (an odd number!) who 'vote', on the basis of the data analysis they have performed, whether or not to continue with the launch. If two vote 'abort' and one votes 'launch', the mission is aborted. Stranger still, some theories of the conscious human mind, pioneered by Marvin Minsky, model it as a system of competing influences that interact rather like 'a society of mind', each 'voting' to determine our actions. Arrow's impossibility theorem applies to all these situations as well. Maybe the next time you feel in 'two minds' you'll know why—especially if you're voting at the time.

Ingram Pinn

Part 5

Simplicity and complexity

We live in strange times. We also live in strange places. The people with whom we populate our universes are the shadows of whole other universes intersecting with our own . . . So give your kid a break, OK?

Extract from '*Practical Parenting in a Fractally Demented Universe*', Douglas Adams, *Mostly Harmless*

Is the world simple or complicated? It depends, of course, on who you ask. Talk to the particle physicist and they will soon be persuading you of the wondrous simplicity of nature, the economy of her forces, the unity behind the diversity. Come back in a year or two and all will be united into a single all-encompassing 'Theory of Everything'. But happen upon a biologist or a neurophysiologist and you will be overwhelmed by the sheer complexity of it all—a higgledy-piggledy mess of competing influences—and we haven't even got to the study of human behaviour yet. At least the biologist can be confident that his predictions about the behaviour of fish won't change the way fish behave, but forecast the result of an election or the trend of an economy and you will alter their outcomes in ways that are unpredictable, in principle.

This problem of the contrast between simplicity and complexity is addressed in this section. It shows how different scientists lay stress upon different aspects of Nature's workings. Some focus upon the laws of Nature; others upon their outcomes. The distinction tells us something important about the nature of the world and reveals the secret of how our Universe can manifest a bottomless well of complex asymmetrical structures yet be governed by a very small number of simple, symmetrical laws. This chapter explains the impact of order and disorder on the workings of the world and the unusual, new 'critical' structures that have been found to emerge on the borderline between the two regimes. Like the constant adjustment of the slope of a sand pile that occurs as sand is added to it, so natural systems can organise themselves into stable large-scale patterns that emerge from a huge number of individually chaotic events. The newly discovered process of 'self-organising criticality' serves as a paradigm that may help us understand the development and persistence of a wide range of complex systems in Nature, from the weather to ecosystems and traffic jams.

Until quite recently the march of physical science has stayed close to the path of studying the regularities of Nature and the simple laws behind them. Our patterns of scientific education are built around them. But during the last 15 years there has been a revolution in our approach to Nature. The advent of powerful, small, inexpensive computers with good interactive graphics has opened up the study of the complex by exploratory methods. A technological innovation has opened a door into a previously intractable realm. Within its domain we have found chaos and organisation that mix chance and necessity in subtle and unsuspected ways. In the first chapter we ponder some profound features of complexity that are displayed by watching a pile of sugar. In the chapters '*Where the wild things are*', '*Patterns emerging from a complex world*' and '*Chance and determination*', some further aspects of these new discoveries are discussed, along with their ramifications for our views about chance and providence. The impact of the concept of 'chaos', or disorganised complexity, has been deep and wide. Although the professionals have shifted their attention to the more interesting realm of organised complexity, the idea of chaos has worked its way into a range of other subjects. It has led to movies like Steven Spielberg's adaptation of Michael Crichton's *Jurassic park* and plays like Tom Stoppard's *Arcadia*. These dramatic uses of chaos have stimulated some writers to search for a new way of analysing the essential features of great literature in the light of unpredictability and sensitivity to the possible outworkings of the events they describe. In '*Outlook moderate to fair; winds, changeable*', I review a book that applies the lessons of chaos—sensitivity of the future to very small changes in the present—to literary criticism. The closing piece, '*Why is the world comprehensible?*' takes these insights about the contrasts between the complexity and simplicity of the world to shed some light on the question of why we can understand so much about the structure of the Universe. What are the key aspects of its structure that render it so amenable to study using relatively simple mathematics: why are the patterns we find in the universe so much simpler and smaller than even we can imagine?

CHAPTER 17

Complexity

The love of complexity without reductionism makes art; the love of complexity with reductionism makes science.

E.O. Wilson

The days seem to have gone when a Galileo or a Newton could make profound discoveries about the workings of Nature and her laws with apparatus assembled from the local market. Pieces of string, prisms of glass, pieces of metal and wood, when suitably interrogated, told these men some of Nature's deepest secrets about light and matter in motion. Today, fundamental science is dominated by vast experiments deep underground or instruments orbiting the Earth in satellites that cost hundreds of millions of dollars. They are supported by small armies of technicians and take a succession of committee meetings just to decide when to switch the apparatus on. Even biology, for so long the preserve of naturalists armed with nothing more complicated than a butterfly net, has moved into the big-science league. The quest for the structure of the genome that will decode the encrypted secrets of life itself is the centrepiece of biological research and involves international teams of scientists, computers, technicians, and a multi-million dollar budget.

On this view, bigger seems always to be better. But, in recent years something profound about Nature's workings has emerged from watching something so simple that you might do it to pass the time at the bar whilst waiting for your late-coming companion to turn up. Slowly pour some salt or sugar vertically down onto a flat surface. Watch as it builds up into a pile. At first, the pile just steepens steadily. Little avalanches keep occurring and move material down the slope. But then something interesting happens. The pile reaches a particular steepness; then it gets no steeper. Avalanches of all sizes keep occurring but serve only to keep the slope poised at this special 'critical' angle. This also happens with grains of rice or sand. The pile may end up in a different shape but the basic pattern of behaviour will be the same. A concatenation of individually unpredictable events results in an impressively ordered heap.

Scientists have realised that something profound can be learned from this simple sequence of events. The individual falling grains have trajectories that are 'chaotic'; that is, a very small change in the way that they are dropped will produce a large change in their subsequent career: they end up in a very different series of places in the pile. When the first grains are dropped on the pile they tend to affect only grains that are close by. But as the pile steepens new additions will produce avalanches of grains whose effects extend over more and more of the side of the pile. When the critical slope is reached an avalanche can extend over the whole of the side of the pile. If we built up our pile on a small plate then when the base of the pile reaches the edge of the plate the grains fall off the edge at the same rate as they are being dropped onto the pile.

This type of process has been given an exotic name, 'self-organising criticality', by the Danish physicist Per Bak who was the first to appreciate its significance from observations of piles of sand. The name captures two aspects of what is going on. First, the pile seems to organise itself into a slope with a special angle of inclination. It does so by a complex collection of avalanches. Second, the pile is always on the verge of an avalanche, yet is in equilibrium. It is steady without being static. In this 'critical' steady state the pile organises itself into an equilibrium that is maintained by the possibility of avalanches of all sizes. If there were a special size for collections of grains to stick together and form avalanches then this self-organising critical behaviour would not occur.

We have begun to appreciate that this mundane experiment with grains of sand or sugar may be telling us how many superficially different forms of complexity develop in Nature. Some have even speculated that it might lie at the root of all examples of complex organisation. How could this be? Instead of avalanches of sugar maintaining a steady slope of our pile, think of earthquakes or volcanic eruptions maintaining a complex balance of pressure in the Earth's crust. Or, we could replace the avalanches by bankruptcies and stock-market crashes contributing to a complex economic balance. Replace them by extinctions of species and we could be studying how a delicate ecological balance is maintained in a natural habitat. Just as the avalanches play a positive role in maintaining the critical slope, so the extinctions may open up new niches for plants and animals that were previously crowded out by others. Another interesting application of this picture has been to the study of traffic flow. How many times have you sat in a traffic jam on a main road for no apparent reason: No accident. No breakdown. Traffic flow seems to stop and start for no obvious reason. What may be happening, through the independent haphazard movements and responses of individual drivers, is that the traffic flow is organising itself into a critical state. Instead of using avalanches of all possible sizes, it uses traffic jams of all sizes. Just as we cannot attribute reorganisations of the sugar pile to the effect of a specific grain when it is in its critical state, so we cannot trace a simple individual cause of the traffic jam that we happen to find ourselves in. This hypersensitivity occurs in the critical state and is just what is required to maintain a complicated balance over a wide range. There is no way you can achieve that by influencing only your nearest neighbours.

Traffic jams of all sizes is an optimal way to use the road space from a global point of view. Lots of space between cars is a big waste of capacity if you are a global organiser. It's infuriating for us when we are stuck in a traffic jam for no reason, but the next time it happens just reflect that this is one way complex systems organise themselves.

The study of how complex systems organise themselves is currently one of the great frontiers of scientific research. It promises to tell us new things about economic systems, ecological balances, weather systems, turbulent liquids, even the workings of the human mind. Our example seems to be telling us things about principles governing the development of complicated equilibria that are more far-reaching than those which govern piles of sand and sugar. Perhaps there are other, equally simple, situations in Nature which could act as models for other complicated things in the world.

One of the most fascinating things about the study of complex structures is the window they may open up on some of the creative arts. The simplest of these is music. The data is precisely and completely recorded and provides a one-dimensional pattern of sounds in time. The results are wonderful examples of organised complexity that have been mysteriously selected by our minds, and are at least as complex as the examples that we find occurring naturally. Strikingly, a few years ago, two American physicists, John Clarke and Richard Voss, discovered that the musics that humans of many cultural traditions have created possess a similar statistical feature. Whether it is the music of the African Pygmies, Indian Ragas, Russian Folk Songs, American rhythm and blues, Beethoven, or the Beatles, it follows a special underlying pattern. If we look more carefully at this special pattern, it seems to be the musical counterpart of the 'critical' slope of our pile of sugar. It is a musical sequence that, on average, contains patterns, what engineers call 'correlations', over all intervals of time.

This situation suggests that the musical compositions that we are attracted to may well be examples of a 'critical' state of self-organisation. Remember that such a state is characterised by its optimal sensitivity to small changes. Perhaps this is one of the things that we appreciate most about pieces of music. It means that they are sensitive to the smallest nuances of performance. Different performances of the same work can then always seem fresh, always have something new for their listeners to experience, and possess an attractive type of unpredictability that spans intervals far longer than those of the individual notes, pauses, and movements which combine to create the whole. Variations at one place or another produce an overall impression. We like them if they help to maintain the overall organisation. Banal music lacks this sensitivity. It is invariant: dead.

So, next time you are stuck in traffic on the way to the concert, or find yourself putting sugar in your coffee at the bar afterwards, spare a thought for the complexities of life.

CHAPTER 18

Where the wild things are

Emergent behaviour, by definition, is what's left after everything else has been explained.

George Dyson

Watchers of science are much impressed by the very large and the very small. Authors, never remiss to fill a vacant evolutionary niche, have populated the shelves of bookshops with their offspring: breathless volumes with catchy titles filled with news about the latest discoveries and speculations from the inner space of elementary particles and the outer space of distant stars and galaxies. But the really wild things live in the in-between world of the not very large and the not very small. Here it is neither the stringy quantum things nor the cosmological exotica that hold sway; rather it is many-sided complexity—pure and simple.

What characterises the everyday world about us and within us is the complicated coming together of many simple things. The way in which these ingredients are organised gives rise to new types of behaviour that are complicated enough to defy the traditional means of scientific analysis and spawn new ones. The human mind is the most complicated natural phenomenon of which we are aware, not entirely unexpectedly for were it significantly simpler, then we would be too simple to know it. In past centuries, it was suspected that life was endowed by some *élan vitale* that distinguished living matter from other matter. This view now lacks any foundation and we have come to appreciate that it appears closer to the truth to regard life and mind as manifestations of a certain level of complexity in the organisation of matter. What matters about the brain or a computer is not the matter of which it is made but the matter of how it is wired together.

But not everything that is complicated is alive. The weather, the behaviour of economies and opinion polls, turbulent liquids and the instabilities of animal populations, all possess complexities that defy us to quantify them. Yet the very

---

A review of *The dreams of reason: the computer and the sciences of complexity*, by Heinz R. Pagels, Bantam, 1990

diversity of such complexity hints that it must be possible to extract the essence of the complexity from each specific application and study it in the abstract, just as we might isolate a piece of computer software from any particular brand of hardware. The essence of the software is the organisation of the logic within it, rather than the laws of electromagnetism or the hardware we choose to run it on. Just as that disembodied logic obeys certain rules, so we might hope to find patterns governing the evolution and nature of all complexity.

Mathematicians are familiar with such grandiose aims and parts of their subject are dedicated to the study of all possible changes subject only to very general restrictions on their qualitative character and the proviso that they be not unrealistically special (like a pin balancing on its point). Heinz Pagels' book is a personal response to this emergence of the study of complexity. Because complexity is everywhere, within us and around us, this is a springboard from which any depths can be plumbed. As a result, the book is a strange mixture of autobiographical reminiscences of his teenage and student experiences in the wackier parts of California, the new science of complexity and his personal reactions to some traditions in the philosophy of science—all made more poignant by the unusual circumstances of the author's death two years ago.

Those who have enjoyed Pagels' matter-of-fact writing which characterises his two previous books of popular science, *The cosmic code* and *Perfect symmetry*, would not guess that they were reading the same author if they meandered through these pages. Yet, although the text is nonlinear and complex in its development, it is far from chaotic. There are many interesting expositions of the new ideas permeating the study of randomness in mathematics and its application to the real world. The reader will learn of the new definitions of algorithmic randomness that characterise a random sequence as one that cannot be abbreviated while retaining the information it contains. That is, there is no simple formula that will encode the information content of the sequence. It is its own shortest representation. Mathematicians describe it as 'algorithmically incompressible'. Indeed, we can see that for some reason our entire Universe possesses the property of being algorithmically compressible. Science is simply the search for those compressions. Instead of filling our libraries with the observations of all the motions that have ever been observed on Earth or in the heavens, we find that simple laws of motion enable us to regenerate them. Our laws of Nature are the algorithmic compressions of the reality available to us. The search for some 'Theory of Everything' is a manifestation of an as yet unsubstantiated belief that there exists some simple set of equations that encode the internal logic of the Universe.

When he comes to a discussion of the philosophy of science, Pagels does not present his own system of thought, merely important factors that he believes to have been neglected by philosophers of science. He is evidently aware that sometimes the human mind sets about making simple sense of complicated reality, effecting its own algorithmic compression, by simply ignoring or discarding the information that it cannot make sense of or abbreviate into simple general principles. However, while I do not doubt that there are some philosophers of science to be found who could be guaranteed to miss any point, I think Pagels is simply not well-enough informed about the diversity of ideas within that subject to make his points good. He simply lumps all its scholars into one homogeneous mass. And many of his interesting discussions are not as new as he evidently thinks them to be. For example, his emphasis upon science as an evolving system in which there is an effective analogue of natural selection that selects for agreement with the facts rather than merely some unguided system of thought will be familiar to many.

Pagels appears to be a realist who believes that it is best for one's intellectual health and that of science in general to be a constructivist. He holds the practical implementation of abstract notions in working devices to be a great antidote to wrong-headedness. Throughout his discussion he lays great stress upon the objectivity and universality of science in contrast to the humanities. He tells an interesting story of his experience at a literary dinner party when conversation turned to reviews in the *New York Review of Books*. Despite his liking for this august journal and his avid reading of many of its reviews, he admits to having no long-term memory for anything in them. He believed his own mind, like that of many scientists, to be tuned to resonate to the invariance rather than the ephemera of experience. No-one spoke to him for the rest of the evening.

This is a difficult book to summarise because it bears many of the marks of an attempted synthesis of all the author's thoughts on a wide spectrum of subjects that do not naturally come together into a seamless whole. Nonetheless, it contains much that is worth reading and pondering. Francisco Goya wrote 'The dreams of reason bring forth monsters', the words that inspire its title. But it shouldn't give you nightmares. It is not an exposition of science. It is not a work of philosophy nor is it an autobiography. But these are three good reasons for reading it.

CHAPTER 19

Patterns emerging from a complex world

The Universe is a computer, the only trouble is that somebody else is using it.

Tom Toffoli

The history of the sciences displays two recognisable types: the unifiers and the intricacists. The unifiers are fascinated by the harmony and simplicity at the root of complicated sequences of events. Among their ranks one finds mathematicians and physicists of a theoretical persuasion who ponder the parts of the world that are little and large. Whereas the unifiers have tried to synthesise the world out of little pieces, the intricacists have sought to understand the whole by studying its pieces bit by bit. Among their ranks one numbers naturalists, psychologists, meteorologists and sociologists, along with all manner of physicists and chemists dedicated to the study of exotic materials in every shape and form. While the second strategy sounds admirable it is liable to descend into an elaborate form of stamp collecting, gathering fact after fact in the hope that there is a grand unifying theory to pull them together. Meanwhile, the unifiers have unravelled simple laws, solved all manner of mysteries about the world and, above all, developed a masterly facility for distilling off the simple essence of a situation so that they can describe its workings approximately. As a result, the unifiers have dominated many of the scientific disciplines.

This has begun to change with the advent of fast, inexpensive, small computers, which have added a formidable weapon to the armoury of the intricacist. Complex systems and sequences of events can now be simulated and watched. Their evolution can be speeded up by enormous factors. Problems which are intractable using exact mathematical methods become the focus of experimental mathematical study, opening a new frontier of fundamental science. What characterises the objects of study is collective behaviour that amounts to

---

A review of *Complexity: the emerging science at the edge of order and chaos*, by M. Mitchell Waldrop, Viking, 1993

more than the sum of its parts. A system like the brain does what it does, not because of the nature of the pieces out of which it is made, but because of the way in which those pieces are organised.

This book tells the story of some of these searchers after complexity: how they germinated ideas about the development of organised complexity and the emergence of transient order in evolving systems far from the simplicities of equilibrium; how they were pulled together by the dream of the Santa Fe Institute in New Mexico; and how their vision of a new interdisciplinary science of evolving complexity has itself evolved in unexpected ways. Mitchell Waldrop has focused on the interesting personalities that have played a role in the development of the study of organised complexity. They are a band of economists, biologists, computer scientists, physicists and mathematicians who thought they had nothing in common until they were forced into interaction by the interdisciplinary programmes that were initiated by the Santa Fe Institute. Set up in 1984 to act as a catalyst for research into complex systems, the institute plays host to a stream of visiting researchers as well as building links with other centres of research.

The style of the book is serious journalism with a strong emphasis on personal conversations to try to convey an inside feel for the style of Santa Fe and the style of work it has encouraged. Waldrop describes some of the scientific projects in great detail, particularly those in the biological area, but is sketchy on others, especially the economics projects which seem to be a major part of the institute's programme. But by focusing on individuals, the story avoids becoming stuck in the groove of just explaining the science of complexity. The price to pay is that the book ends up being rather too long, with repetition of general scientific ideas. Indeed, the author seems to have begun to weary himself, because the book peters out into an unmemorable end. But along the way there are some nice lines from the protagonists: the University of Arizona—'a real Ken and Barbie kind of place'—or the image of a mountaineer about to descend into a hole in the ground, 'because it's not there'.

A recurring lesson from Waldrop's story is how complex systems are providing an interesting antidote for philosophers of science who place too much emphasis upon falsification and prediction as the hallmarks of science. When faced with a complex system such as the Earth's weather system or an economy, we may have mathematical models of all or part of it. But because the system will exhibit extreme sensitivity to its starting conditions we will not be able to predict the future with that model. We can predict aspects of the weather, but not the detailed whole. However, one can understand and explain any atmospheric phenomena that are seen by using the ingredients of the model. The comprehension and explanation of observed trends is just as fundamental to the study of complex systems than the ability to predict the future.

The students of organised complexity are seeking general principles that can specify what complexity is and determine how it develops. If such generalities exist then they would have application across a wide range of disciplines. At present the search for such principles is pursued through thorough investigation of subjects which display novel types of complexity. In this way one builds up a repertoire of examples and ideas which can be stripped of their specific applications in, say, biology or economics, in order to isolate the general lessons that underpin them.

Waldrop has done a good job of conveying the intellectual atmosphere surrounding the emergence of the study of complexity. This is not the book to read if you want to learn about complexity itself, for while the author provides quite a lot of background discussion of the problems it tackles, that is not his primary aim. He has succeeded in showing that the social organisation associated with a collective intellectual development is itself a complex system which exhibits many of the vital features of emergence, modified adaption, random drift and selection that typify many of its objects of study.

CHAPTER 20

Chance and determinism

I feel like a fugitive from the law of averages.

Bill Maudlin

The ancient cultivation of the notion of lawfulness in Nature led to a categorisation of chance as something to be shunned or revered. Chaos and irrationality represented the fate of those who spurned the gods. Yet chance could be the mouthpiece of the gods. The drawing of lots was the means by which the word of God was revealed to the Jews and the high priest carried with him a random device for producing answers of yes, no, or wait, when Yahweh was consulted.[1] By the standards of other areas of mathematics and science, probability took a very long time to emerge as a branch of quantitative study.[1,2] This might have been because of the awesome religious implications of quantifying the voting patterns of the gods. Even today, English law recognises natural disasters as 'acts of God'. Yet we know that most ancient cultures gave rise to gambling and games of chance. But those games lacked a general aspect that would enable universal principles and 'laws' to be devised for them. Instead of symmetrical dice, each of whose outcomes is equally probable, the ancient gambler would use a knuckle bone or some other asymmetrical object whose different possible orientations after being thrown were not equally likely. There could be a quantitative pattern or theory of the behaviour of each one of these devices but none that would apply universally to all. That generalisation requires the simplification that arises from the notion of equally likely outcomes.

The close association between the idea of laws of Nature and the monotheistic religious traditions of the West has long been seen as an important factor in providing a body of ideas conducive to the development of science.[3] The notion of laws of Nature as the decrees of a divine lawgiver helped to underwrite the universality and the unchangeability of those laws. It also served to perpetuate the ancient association of chance and apparent lawlessness with irrationality and opposition to the divine purpose. Order and chaos were natural enemies.

Until the end of the 19th century, physicists focused their attentions entirely upon the laws and regularities of Nature. They studied the simple, symmetrical, and soluble problems of mathematical physics which yielded to analysis by pencil and paper. The first person to hint at something other than the lawful predictability of the laws of Nature that Newton had revealed was James Clerk Maxwell. In February 1873, Maxwell was asked to deliver an evening discourse on the subject of free will and determinism. His audience of Cambridge colleagues expected him to discuss the dilemmas for the issue of human freedom introduced by the recently announced theory of evolution. Instead, Maxwell talked about something totally unexpected that grew out of his studies of gases of molecules. He pointed out that while the laws of motion laid down by Newton display predictability in principle (that is the future state of the world is uniquely and completely determined by its present state) they do not lead to determinism in practice.[4] This is because we can never determine the present state of the world with perfect accuracy, and many natural processes lead to a rapid amplification of that initial uncertainty so that after a short time we know essentially nothing about the true state of the system. Maxwell himself gave the example of a point on a railway line to illustrate sensitive dependence on small changes. A slight alteration of the point's switch leads to a divergence from the track that the train would otherwise have followed.

These systems, which exhibit a sensitive dependence upon their starting states, are now called 'chaotic systems' and recently they have been much in the news.[5] The publicity surrounding them has led to a widespread popular belief in the ubiquity of chaos and randomness. Two simple examples illustrate the notion. Weather systems are chaotically unpredictable, because we are unable to determine the present state of the atmosphere at every place. We have weather stations every few miles over built-up areas but they are far sparser over the oceans. The variation in the state of the weather that is possible in between the weather stations is sufficient to lead to very different future states for the weather everywhere, no matter how accurately we measure its characteristics at the weather stations. A more spectacular example is provided by the cue games of snooker, billiards, or pool. We know from experience that these games exhibit sensitivity to their starting conditions because the slightest miscue produces a large error in the trajectory of the ball. Suppose we were playing in idealised conditions so that there was no air resistance or friction with the table to slow the balls down. And suppose also that we could strike the ball as accurately as Heisenberg's Uncertainty Principle of quantum mechanics allows (this allows accuracy far greater than the radius of a single atomic nucleus and is completely unrealistic of course). Now, even with all this artificial precision our knowledge of Newton's laws of mechanics would be quite useless to us after the ball had made only 10 successive collisions with other balls and the sides of the table. This is all it takes for the initial quantum uncertainty in the location of the ball to grow larger than the size of the table. Thus, determinism in practice is quite different to determinism in principle. Most complex systems exhibit this sensitive dependence on their starting conditions. A notable example is the evolutionary process. A small change in genetic variation in the past will lead to huge variation from what would have ensued had it not occurred. Isaac Asimov's 'psychohistorian' Harry Seldon, who appears in *The foundation* science fiction trilogy,[6] cannot operate because of the rapid future divergence of presently similar mental states.

Until very recently scientists had largely ignored the study of chaotic and complex systems because they are extremely difficult to study. We recognise that the teaching of science has crystallized around the exposition of soluble problems. And by 'soluble' one often means soluble with pen and paper (and pure thought) in 30 minutes in an examination hall. While there is some scope to be more adventurous than this in advanced courses, this pattern still lies at the foundation of the teaching of the core material in degree courses like physics. But this pedagogical inertia is only part of the story. While subjects like chaos and complexity are exciting and vital to the understanding of all manner of real world problems, they are not well suited to exposition within the framework of traditional teaching structures. Why? By definition, chaotically unpredictable physical systems are not simple exact solutions of the equations of physics. Their study requires the art of computer modelling. One seeks insight as much as a mathematical solution of the problem in hand. To achieve this one simulates the evolution of complex systems involving many interacting parts using computers[7] and studies the outcomes observationally using carefully designed graphics. Indeed, it is no coincidence that the entire study of complexity and chaos has grown up in tandem with the advent of the inexpensive personal computer. It is experimental mathematics.[8]

There is a further level of unpredictability that sensitive systems like the weather do not display. Forecasting the weather does not lead to any change in the weather. But forecasting the course of the economy or the result of an election can lead to changes in the economy and can alter the outcome of the election in ways that are unpredictable in principle as well as in practice.

The most interesting feature of so many natural systems, like the motion of the molecules in the air around us, is that they exhibit behaviour that is stable on the average over a long period, despite being composed of many chaotic microscopic processes. This is often a consequence of what statisticians call 'the law of large numbers'. When a very large number of *independent* events occur they will follow a general statistical pattern called the normal, or Gaussian, distribution. Thus, while one cannot predict when an individual will die, actuaries produce very accurate predictions of the average life expectancies of large groups of individuals in such a way that insurance companies make a healthy profit from their premiums. Of course, this breaks down if some unforseen event (like asbestos poisoning) produces a widespread series of disasters that are not independent.

All that we have been saying applies to 'classical' physics. Quantum unpredictability plays no role in this concept of chaos. It enters only to make it inevitable that we can never determine the state of any system with perfect accuracy. So far, the study of how classically chaotic systems behave when studied in the context of quantum theory remains an unsolved problem. Quantum changes are less sensitive than non-quantum ones but quantum systems change in sudden and uncertain ways when they are observed.

These developments in the study of behaviours that appear 'random' because they are extremely complicated and practically impossible to link causally back to specific starting states have provoked a careful study of what is meant by randomness and order. If we are shown a string of numbers then we will say that it is non-random if we can see some pattern in it, while we call it random if no pattern can be discerned. This simple practice leads to a more formal definition due originally to Kolmogorov[9] and Chaitin.[10]

If the information in a string of symbols can be compressed into a programme or formula that is shorter than the sequence itself then we say that the sequence is non-random: it is compressible. But if no such abbreviation of the sequence exists then it is random or incompressible. If you want to tell someone else what is in an incompressible sequence then you can only send them the listing of the sequence, but if the sequence is compressible you could send them the abbreviated formula. The length of the shortest programme that will generate the sequence is called its algorithmic complexity. Whilst algorithmic complexity captures some aspects of randomness and order it is not entirely satisfactory because it attributes too much complexity to random noise. If lots of monkeys were seated at typewriters to generate a 'book' then what they produced would be almost pure random noise and would therefore have a large algorithmic complexity. By contrast, the works of Dante would have a smaller algorithmic complexity but we sense that they are not in any sense simpler than the monkeys' efforts. To capture this distinction the notion of 'depth' has been introduced which gives the amount of entropy (or heat) that is produced by the shortest programmes producing the sequence.[11,12] This takes note of the fact that if a great deal of effort must be made to produce a compression then the object being compressed must be deeper and more complex than if the compression is obvious. By this criterion the works of Dante are logically deep while the output of the monkeys is not.

It is interesting to speculate about the types of order and non-randomness that we like. At the beginning of this century George Santayana delivered a famous series of lectures on aesthetics entitled 'The sense of beauty'[13] in which he recalled the fascination that humanity has always had for the appearance of the night sky. He thinks that its attraction to the ancients was because it manifests a level of complexity that is intermediate between unfathomable complexity and trivial simplicity. Its patterns are both challengeable and challenging. We see this situation with a popular game like the Rubik cube or John Conway's 'Game of life'.[14] Were it significantly simpler then it would soon become as uninteresting as noughts and crosses, but if it were much harder then it would become so close to being practically impossible that we would soon lose interest in it. The evolutionary process has found it optimal to allocate a particular budget of resources to pattern recognition. If we could recognise no patterns we could not store information about the environment and adapt to it. If we stored every microscopic detail about the structure of the world around us we would be wasting resources upon unnecessary information storage and our survival would again be threatened.

It is important to stress that just because we can trace some of the historical and social influences that have led to our concepts of natural laws and randomness this does not mean that there is nothing more to these scientific ideas than sociological influences. This is an error made by many sociologists of science who focus upon the often *ad hoc* or irrational ways in which new scientific theories come to be proposed. While the process of theory creation has few rules, and those are often broken, the process by which those theoretical proposals are tested is strict and rigorous. Moreover, it makes no more sense to say (as some do, see ref.[15]) that W and Z bosons were created by the scientific enterprise that searched for them rather than discovered by Rubbia and his colleagues at CERN, than it does to say that Columbus invented America. Mathematical theories predict the masses and properties of these particles with great precision. Moreover, the power of mathematical prediction is most impressive in those areas that are most distant from everyday life. It is the outer space of astronomy and cosmology and the inner space of elementary particle physics where our ability to predict the course of Nature on the assumption of simple laws is most successful. These successful pictures of the Universe are not subject to direct action by natural selection. Our picture of the world on the everyday scale is influenced by the evolutionary process and gives us a faithful image of the nature of true reality in those areas where such an image is a necessary prerequisite for our survival. No such imperative requires our minds to embody accurate images of the nature of the astronomical and elementary particle worlds. How curious that mathematics should so faithfully represent worlds we have never been in contact with since time began.

References

[1] J.D. Barrow, *Theories of everything*, Oxford University Press, Oxford (1991).
[2] I. Hacking, *The emergence of probability*, Cambridge University Press, Cambridge (1975).
[3] J.D. Barrow, *The world within the world*, Oxford University Press, Oxford (1988).
[4] J.C. Maxwell, in *The life of James Clerk Maxwell*, eds L. Campbell and W. Garnett, London, Macmillan (1882).
[5] J. Gleick, *Chaos: making a new science*, Viking, New York (1987).
[6] I. Asimov, *Foundation*, Grafton, London (1988). First publ. (1942).
[7] L. Smarr, *Scientific computing*, Scientific American Library, Freeman, San Francisco (1993).
[8] J.D. Barrow, *Pi in the sky: counting, thinking and being*, Oxford University Press, Oxford (1992).
[9] A.N. Kolmogorov, *Inform. Trans*, 1, 3 (1965).
[10] G.J. Chaitin, Scientific American, (July 1980), pp. 80–5. *Algorithmic information theory*, Cambridge University Press, Cambridge (1988).
[11] C.H. Bennett, 'The thermodynamics of computation: a review', *Int. J. Theor. Phys.* **21**, 905 (1982).
[12] H. Pagels and S. Lloyd, 'Complexity as thermodynamic depth', *Ann. Phys.* **188**, 185 (1988).
[13] G. Santayana, *The sense of beauty*, Dover, New York (1955).
[14] W. Poundstone, *The recursive universe*, Morrow, New York (1985).
[15] A. Pickering, *Inventing quarks*, Edinburgh University Press, Edinburgh (1986).

CHAPTER 21

Outlook moderate to fair; winds, changeable

Let chaos storm!
Let cloud shapes swarm!
I wait for form.

Robert Frost

This is one of those books whose non-existence would require its invention. It latches on to the contemporary fixation with the subject of chaos following the exhaustive popularisation of its role in science over the past decade and uses it for an experiment in literary criticism. 'Band wagons roll' I hear you cry. However, this book is by no means the sort of trivial importation of the scientific that many social studies have perfected. The author focuses on a number of recurring stories that involve a significant bifurcation in human affairs at some stage: a choice that has incalculable consequences.

The author's favourite examples are *Paradise Lost*, *Hamlet*, *Macbeth*, *The Tempest*, and many modern retellings of these stories. In tracking this unpredictable market of literary derivatives one is bounced around an arcade of creations that span a bewildering range of times, media and genres: from *Star trek* episodes, *Jurassic Park* (the book and the film), *Arcadia*, *The Island of Doctor Moreau*, *Perelandra*, *The Forbidden Planet*, and many others. The author's aim is to track the sensitive development of narratives containing critical choices, eating or ignoring the forbidden fruit, for instance, while expounding a view of the complexity of literary construction that reflects the multiplicity of possible interpretations that a given work presents at diverse times, to different readers, in varied contexts. Hawkins's account of a variety of familiar books, films and works of drama is often illuminating and refreshingly clear of unnecessary complications. By concentrating on well-known works rather than drawing from a library of esoterica, the author has managed to produce a story that will be

---

A review of *Strange attractors: literature, culture and chaos theory*, by Harriet Hawkins, Prentice Hall, 1996

rewarding light reading for scientists as well as for literary professionals. Yet, for all the author's skill in painting a picture of chaos in the mind of the reader, the scientist might be left with a nagging doubt that it is not ultimately of any real use.

Chaos is not merely a way of looking at the world. Its development occurred because of a desire to describe particular complicated phenomena, like turbulent fluid flow or changing weather patterns, in a precise fashion. While one can apply chaos theory to texts, just as one can apply mathematics to anything, there is no guarantee that it will amount to more than a notation. Thus, although I believe the author's general thesis to be correct, can it really provide literary critics with the depth of insight that they seek? For the whole approach to chaos employed by mathematicians is something of a Faustian bargain. Detail is sacrificed on the altar of generality.

Chaotic systems are critically sensitive to perturbations; exact descriptions of their behaviour are therefore impossible to create or confirm; so one retreats, to ask if there are common properties of wide classes of phenomena, regardless of their detailed specification. If so, then one can make predictions about their behaviour without needing to model them precisely and completely. We pay a price because if we can discover general properties of whole classes of chaotic phenomena irrespective of their detailed character, then that information must be rather weak. Apply the same approach to literary constructions and we end up with statements of such generality but low specificity that they are little more than the painful elaboration of the obvious.

Artistic criticism, whether it be of painting or writing is not the exposition of a general theory in the way that scientific theory is. It possesses intrinsic unpredictability which derives from the uniqueness of the individual doing the commentating. There is no more reason to expect new literary insights from the application of chaos theory to literary texts than to expect that the Meteorological Office will discover new structures in weather systems from a careful study of *The Tempest*.

Finally, one should alert chaotic critics to a dramatic irony. During the period that literary critics have taken on board the concept of disordered chaotic complexity, the professional chaologists have moved on. Recent studies of complex systems have focused upon examples of organised (rather than disorganised) complexity. Here, I believe there is a closer point of contact with artistic creations. Musical or literary compositions are more fruitfully viewed as particular examples of organised complexity in the large than as stories about the sensitive dependence of human affairs. Perhaps Hawkins can be encouraged to take up the new challenges offered by organised complexity and its delicate interface with chaos in an equally charming and entertaining fashion.

CHAPTER 22

Why is the world comprehensible?

Can we actually "know" the universe? My God, it's hard enough finding your way around in Chinatown.

Woody Allen

One of the most striking features of the world is that its laws seem simple whilst the plethora of states and situations it manifests are extraordinarily complicated. To reconcile this disparity, we must redraw a sharp distinction between the laws of Nature and the outcomes of those laws, between the equations of physics and their solutions. The fact that the outcomes of the laws of Nature need not possess the symmetries of the underlying laws makes science difficult and teaches us why the complicated collective structures we find in Nature can be the outcomes of very simple laws of change and invariance. But, however necessary an appreciation of this point might be, it is far from sufficient to make sense of the physical world.

On the face of it, we might imagine that it would be far easier for the world to be an unintelligible chaos than the relatively coherent cosmos that the scientist delights in unravelling. What are the features of the world which play important roles in rendering it intelligible to us? Here is a catalogue of those aspects which appear to play a subtle but vital role in the intelligibility of Nature.

Linearity

Linear problems are easy problems. They are those problems where the sum or difference of any two particular solutions is also a solution. If L is a linear operation and its action upon a quantity A produces the result a, whilst its operation upon B produces the result b, then the result of the operation of L upon A plus B will be a plus b. Thus, if a situation is linear or dominated by influences that are linear, it will be possible to piece together a picture of its whole behaviour by examining it in small pieces. The whole will be composed of the sum of its parts. Fortunately for the physicist, a large part of the world is linear in this sense. In this part of the world, one can make small errors in determining the behaviour of things at one time and those errors will only be amplified very slowly as the world changes in time. Linear phenomena are thus amenable to very accurate mathematical modelling. The output of a linear operation varies steadily and smoothly with any change in its input. Non-linear problems are none of these things. They amplify errors so rapidly that an infinitesimal uncertainty in the present state of the system can render any future prediction of its state worthless after a very short period of time. Their outputs respond in discontinuous and unpredictable ways to very small changes in their inputs. Particular local behaviours cannot be added together to build up a global one: an holistic approach is required in which the system is considered as a whole. We are familiar with many complicated problems of this sort: the surge of water from a tap, the development of a complex economy, human societies, the behaviour of weather systems—their whole is more than the sum of their parts. Yet our education and intuition is dominated by the linear examples because they are simple. Educators display the solutions of linear equations and textbook writers present the study of linear phenomena because they are the only examples that can be solved easily: the only phenomena that admit of a ready and complete understanding. Many social scientists who seek mathematical models of social behaviour invariably look to linear models because they are the simplest and the only sort they have been taught about. Yet the simplest imaginable non-linear equations exhibit behaviour of unsuspected depth and subtlety which is, for all practical purposes, completely unpredictable.

Despite the ubiquity of non-linearity and complexity, the fundamental laws of Nature often give rise to phenomena that are linear. Thus, if we have a physical phenomenon that can be described by the action of a mathematical operation f upon an input x, which we denote by $f(x)$, then in general we can express this as a series of the form

$$f(x) = f_0 + xf_1 + x^2f_2 + \dots,$$

where the series could go on forever. If $f(x)$ is a linear phenomenon, then it can be very accurately approximated by the first two terms of the series on the right-hand side of the equation; the remaining terms are either all zero or else they diminish in size so rapidly as one goes from one term to the next that their contribution is negligible. Fortunately, most physical phenomena possess this property. It is crucial in rendering the world intelligible to us and it is closely associated with other aspects of reality, the most notable of which we shall highlight next.

Locality

The hallmark of the entire non-quantum world is that things occurring here and now are directly caused by events that occurred immediately nearby in space and time: this property we term 'locality' to reflect the fact that it is the most local events that exert the predominant influence upon us. Usually, linearity is necessary for a law of Nature to possess this property, although linearity is not sufficient to guarantee it. The fundamental forces of Nature, like gravity, diminish in strength with increasing distance away from their source at a rate that ensures that the total effect at any point is dominated by the nearby sources rather than those on the other side of the Universe. Were the situation reversed, then the world would be erratically dominated by imperceptible influences at the farthest reaches of the Universe and our chances of beginning to understand it rendered pretty slender. Interestingly, the number of dimensions of space which we experience plays an important role in ensuring this state of affairs. It also ensures that wave phenomena behave in a coherent fashion. Were there four dimensions of space, then simple waves would not travel at one speed in free space, and hence we would simultaneously receive waves that were emitted at different times. Moreover, in any world but one having three large dimensions of space, free waves would become distorted as they travelled. Such reverberation and distortion would render any high-fidelity signalling impossible. Since so much of the physical Universe, from brain waves to quantum waves, relies upon travelling waves we appreciate the key role played by the dimensionality of our space in rendering its contents intelligible to us.

Not every natural phenomenon possesses the property of locality. When we look at the quantum world of elementary particles, we discover that the world is non-local. This is the import of Bell's famous theorem. It reveals to us something of the ambiguity between observer and observed that arises when we enter the quantum world of the very small, where the influence of the act of observation upon what is being observed is invariably significant. In our everyday experience, this quantum ambiguity is never evident. We confidently hold such notions like position or speed to be well defined, unambiguous, and independent of who is using them. But the fact that our present-day Universe admits such definiteness is something of a mystery. As we look way back into the first instants of the Big Bang, we find a fully quantum world (see Part 8). From that state, where like effects do not follow from like causes, there must somehow emerge a world resembling our own, where the results of most observations are definite. This is by no means inevitable and may require the Universe to have emerged from a rather special primeval state.

The local–global connection

The helpful presence of linearity and locality in the world of everyday observation and experience was essential for the beginning of our understanding of the world. Such an understanding begins locally and finds local causes for local effects. But what must the world be like in order that we can piece together a global description from the local. In some sense, the global picture of the Universe must be built out of many copies of its local structure. Equivalently, there must exist some invariances of the world as we change the locations in space and in time of all its most elementary entities so that the most basic fabric of reality is universal rather than dependent upon parochial things. Particle physicists have discovered that the world is mysteriously structured in such a fashion and the local gauge theories that particle physicists employ bear witness to the power of this local–global connection. The requirement that there exists this natural correspondence between the local and the global structure is found to require the existence of the forces of Nature that we see. We do not mean this in a teleological sense. It is merely a reflection of the interwoven consistency and economy of the natural structure of things. The forces of Nature are not required as an *ad hoc* ingredient in addition to symmetry.

When we look in greater depth at the mathematical structures that are most effective in the description of the world, and indeed why there can be such effective structures, we find a situation of great subtlety. We find the presence of mathematical operations, like the expansion of a function in the series shown above or the 'implicit function theorem', which guarantees that, if some quantity is completely determined by the values of two variables x and y and is found to be a constant, then y can always be expressed as some function of the variable x alone. These two mathematical properties both define restrictions upon what local information about the world can be deduced from global (or large-scale) information. By applying these local restrictions to themselves over and over again in an iterative manner, we can build up increasingly global information about a mathematical world. By contrast, there exist examples of the converse. Stokes' famous integral theorem and the process of analytic continuation, familiar to undergraduates, are both examples wherein restrictions are defined for the transit from local to global information. They exemplify one of the goals of the human investigation of Nature: to extend our knowledge of the world from the local domain, to which we have direct access, to the wider scale, of which we are as yet ignorant. Stokes' theorem alone does not permit such an extension to be made unambiguously. It leaves an undetermined constant quantity undetermined at the end of the extension process. The power of gauge theories in physics derives from their ability to remove this arbitrariness in the extension process and determine the unknown constant uniquely through the imposition of symmetry and invariance upon the extension process.

All the best physical theories are associated with equations which allow the continuation of data defined at present into the future, and hence allow prediction. But this situation requires space and time to possess a rather particular type of mathematical property which we shall call 'natural structure'. Other theories, like those describing statistical or probabilistic outcomes, which attempt to use mathematics for prediction, often fail to possess a mathematical substratum with a 'natural structure' of this sort, and so there is no guarantee that its future states are smooth continuations of its present ones.

One feature of the elementary particle world, which is totally unexpected when compared with our experience of everyday things, is the fact that elementary particles come in populations of universally identical particles. Every electron that we have encountered, whether it comes from outer space or a laboratory experiment, is found to be identical. All have the same electric charge, the same spin, and the same mass, to the accuracy of measurement. They all behave in the same way in interaction with other particles. Nor is this fidelity confined to electrons: it extends to all the populations of elementary particles, from quarks and leptons to the exchange particles that mediate the four fundamental forces of Nature. We do not know why particles are identical in this way. We could imagine a world in which electrons were like footballs—every one slightly different to all the others. The result would be an unintelligible world.

In fact, even in a world populated by collections of identical elementary particles, there would not exist populations of identical larger systems, composed of systems of those particles unless energy is quantized in some way. Although the uncertainties introduced by the quantum picture of reality are often stressed, this same quantum structure is absolutely vital for the stability, consistency, and intelligibility of the physical world. In a Newtonian world, all physical quantities, like energy and spin, can take on *any* values whatsoever. They range over the entire continuum of numbers. Hence, if one were to form a 'Newtonian hydrogen atom' by setting an electron in circular orbit around a single proton then the electron could move in a closed orbit of any radius because it could possess any orbital speed. As a result, every pair of electrons and protons that came together would be different. The electrons would find themselves in some randomly different orbit. The chemical properties of each of the atoms would be different and their sizes would be different. Even if one were to create an initial population in which the electrons' speeds were the same and the radii of their orbits identical, they would each drift away from their starting state in differing ways as they suffered the buffetings of radiation and other particles. There could not exist a well-defined element called hydrogen with universal properties, even if there existed universal populations of identical electrons and protons. Quantum mechanics shows us why there are identical collective structures. The quantization of energy allows it to come only in discrete packets, and so when an electron and a proton come together there is a single state for them to reside in. The same configuration arises for every pair of electrons and protons that you care to choose. This universal state is what we call the hydrogen atom. Moreover, once it exists, its properties do not drift because of the plethora of tiny perturbations from other particles. In order to change the orbit of the electron around the proton, it has to be hit by a sizeable perturbation that is sufficient to change its energy by a whole quantum jump. Thus the quantization of energy lies at the root of the repeatability of structure in the physical world and the high fidelity of all identical phenomena in the atomic world. Without the quantum ambiguity of the microscopic world the macroscopic world would not be intelligible, nor indeed would there be intelligences to take cognizance of any such totally heterodox non-quantum reality.

Symmetries are small

The possibilities open to an elementary particle of Nature amount to everything compatible with the maintenance of some symmetry. The preservation of some global pattern in the face of all the local freedoms to change is equivalent to a conservation law of Nature and all laws of change can be re-expressed in terms of the invariance of some quantity. The particular patterns that are generated arise from the concatenation of a finite number of ingredients. For example, a collection of patterns might be created from a combination of rotations and straight-line movements in space. The greater the number of distinct operations, or generators, that comprise the overall collection of patterns that can be generated, the greater will be the number of patterns. If this number is very large, then for all practical purposes there will be no discernible symmetry at all. The generators of the symmetries that dictate the interactions that can occur between elementary particles are equivalent to the particles that mediate the force of Nature in question. Hence, the intelligibility of the world relies upon the fact that there are relatively few types of elementary particle. They are numbered in tens rather than in thousands or millions.

There exists one further connection between the population of the elementary particle world and the overall simplicity of Nature. The unification of the forces of Nature relies upon the property of 'asymptotic freedom' that is manifested by the strong force between particles like quarks and gluons which carry the colour charge. This means that, as the energy of the interaction between the particles increases, the strength of their interaction decreases, so that 'asymptotically' there would remain no interaction at all and the particles would be free.

It is this property which allows the disparate forces of Nature that we witness at low energies to become unified at high energies. However, this feature would not arise if there existed too many elementary particles. For example, if there were eight types of neutrino rather than the three that experiments tell us there are, then interactions would get stronger rather than weaker as we go to higher energies, and the world would become intractably complicated as we scrutinized ever finer and finer dimensions of the microscopic world.

The list of properties which may be necessary for the intelligibility of the world is not intended to be exhaustive, merely illustrative. It will not have escaped the reader's attention that many of the properties we have unveiled are also likely to be necessary properties for the existence of complex stable systems in the Universe, a subset of which we would call 'living'. We can conceive of Universes where living observers (not necessarily resembling ourselves) could not exist and, perhaps unexpectedly, we find that there is an intimate connection between the most basic elements of the Universe's fabric and the conditions required for the evolution of life to have a probability that is distinguishable from zero.

We have learnt that it is natural to describe a sequence as random if there exists no possible compression of its information content. Moreover, it is impossible in principle to prove that a given sequence is random, although it is clearly possible to demonstrate that it is non-random simply by finding a compression. Thus it will never be possible for us to prove that the sum total of information contained in all the laws of Nature might not be expressible in some more succinct form, which we shall refer to as the 'Secret of the Universe'. Of course, there may exist no such secret, and even if there does its information content may be buried very deep so that it takes a vast (or even infinite) amount of time to extract useful information from it by computation.

The question of the existence of a 'Secret of the Universe' amounts to discovering whether there is some deep principle from which all other knowledge of the physical world follows. A slightly weaker 'secret' would be the single proposition from which the largest amount of information follows. It is interesting to speculate as to the possible form of such a proposition. Would it be what philosophers call an 'analytic' statement or a 'synthetic' one? An analytic statement requires us to analyse the statement alone in order to ascertain its truth. An example is 'all bachelors are unmarried'. It is clearly a necessary truth, a consequence of logic alone. Synthetic statements are meaningful statements which are not analytic. The physical theories that we employ to understand the Universe are always synthetic. They tell us things that can only be checked by looking at the world. They are not logically necessary. They assert something about the world, whereas analytic statements do not. Some seekers after the 'Theory of Everything' would seem to be hoping that the uniqueness and completeness of some particular mathematical theory will make it the only logically consistent description of the world and this will transform it from being a synthetic to an analytic statement. However, if we want the 'Secret of the Universe' to have testable predictions, it must be a synthetic statement. Yet this is not an entirely satisfactory conclusion because our 'secret' must then contain some ingredients that need to be deduced from some more fundamental principle, and so it cannot be *the* secret of the structure of the entire Universe: for it possesses ingredients that require further explanation by some deeper principle.

This dilemma extends to the problem of the role of mathematics in physics. If all mathematical statements are analytic—tautological consequences of some set of rules and axioms—then we are faced with trying to obtain synthetic statements about the world from purely analytic mathematical statements. In practice, if initial conditions remain unspecified by some form of self-consistency, then they supply a synthetic element that must be added to any analytic mathematical structure defined by differential equations. Even schemes like the 'no-boundary' condition simply introduce certain new 'laws' of physics as axioms.

What is it that makes necessary truths necessary? Presumably it is the feature that they are knowable *a priori*. If we have to carry out some act of observation to see whether a statement is true, then we can only know its truth *a posteriori*. A famous philosophical issue is that of whether all *a priori* statements are analytic. Most of the statements that we encounter in life are either synthetic *a posteriori* or analytic *a priori*. But are there non-analytic statements about the world which have real information content and which are knowable *a priori*? Is a synthetic *a priori* really possible? The most awkward problem would now appear to be how we could know that such a statement was giving us non-trivial information about the world without making some new observation to check. Traditionally, philosophical empiricists have maintained that synthetic *a priori* truths cannot exist, whilst rationalists have maintained that they do, although they have not been able to agree as to what they are. Ever since Immanuel Kant introduced this distinction between analytic and synthetic statements, there have been candidates for a synthetic *a priori* that have since been dispatched to oblivion, statements like 'parallel lines never meet' or 'every event has a cause', which were proposed before the advent of non-Euclidean geometry and the quantum theory.

How then can we have some form of synthetic *a priori* knowledge about the Universe? Kant suggested that the human mind is constructed in such a way that it naturally grasps some synthetic *a priori* aspects of the world. Whereas the real world possesses unimaginable features, our minds naturally sift out certain aspects of reality as though we were wearing rose-coloured spectacles. Our minds will only capture certain aspects of the world and this knowledge is thus synthetic and *a priori*. For it is an *a priori* truth that we will never understand anything that does not register in our particular mental categories. Hence, for us, there are certain necessary truths about the observable world. We might hope to flesh out this type of idea in a different way by considering the fact that there have been found to exist necessary cosmological conditions for the existence of observers in the Universe. These 'anthropic' conditions point us toward certain properties that the Universe must possess *a priori*, but which are non-trivial enough to be counted as synthetic. The synthetic *a priori* begins to look like the requirement that every knowable physical principle that forms part of the 'Secret of the Universe' must not forbid the possibility of our knowing it. The Universe is a member of the collection of mathematical concepts; but only those concepts with complexity sufficient to contain sub-programs which can represent 'observers' will be actualized in physical reality.

Part 6

Aesthetics

That was when I saw the pendulum.

The sphere, hanging from a long wire set into the ceiling of the choir, swayed back and forth with isochronal majesty. I knew—but anyone could have sensed it in the magic of that serene breathing—that the period was governed by the square root of the length of the wire and by π, that number which, however irrational to sublunar minds, through a higher rationality binds the circumference and diameter of all possible circles. The time it took the sphere to swing from end was determined by an arcane conspiracy between the most timeless of measures: the singularity of the point of suspension, the duality of the plane's dimensions, the triadic beginning of π, the secret quadratic nature of the root, and the unnumbered perfection of the circle itself.

Umberto Eco, *Foucault's Pendulum*

Art and science are usually contrasted as polar opposites: one subjective, the other objective; one inventive the other explorative. But offer these opposites to the right bunch of commentators and there will be a long argument as to which is which. There is not much to be gained by stressing the differences between art and science. By focusing upon their areas of overlap and the ways in which they are complementary we may be able to enrich them both.

In recent years, scientists have become deeply impressed by the examples of organised (and even self-organising) complexity they see around them. We saw something of this fascination in the last section. To my mind the creative arts, most notably music, offer wonderful examples of works of intricately organised complexity which are the results of a concatenation of mental events of unfathomable complexity. Science has much to learn from the aesthetic complexities that attract us to works of art. Their attraction resonates with a sensitivity for something more basic that once had direct survival value for us. The creative arts offer our senses some of the most intricate patterns, whereas subjects like mathematics, superstring theory, or structural chemistry can be regarded as a form of highly abstract art. The creative arts may also find that the sciences of the complexity have new insights and images to offer.

Painting has too much going on to be amenable to a direct frontal attack. But a different approach, incorporating insights from evolutionary biology can be illuminating. In the first Chapter, '*Survival of the artiest*', I describe how it may have come about that we like certain landscape scenes by virtue of the advantage such an appreciation provided our ancestors with millions of years ago. An attraction for certain types of landscape, those offering safety, good vantage points, food, and water, was likely to lead to their survival and multiplication with greater probability than those who lack that attraction. Armed with this insight we can begin to appreciate where some of our untutored impulses towards artistic representations spring from. Of course, these impulses can be overwritten by education, fashions, or the appeal of other modern influences.

This approach offers hints as to why we like many things that seem ill-fitted to our current situation. The hotchpotch of genes out of which parts of our aesthetic preferences emerged, at first as mere by-products, was laid down long ago in landscapes more like those of the African savannah than those of the modern world. The design of our parks and gardens betrays our vestigial liking for such landscapes: open spaces interspersed by refuges, trees, bushes, and other sites from which one can see without being seen. We also learn something deeper and wider about the nature of art by thinking in this way. Many artistic representations offer us safe ways to experience new and unusual things, whether they are sights or sounds. Perhaps it was this aspect of our creative imaginations that first offered us such an advantage over other species. Like the horror movie or the roller coaster, it offers a safe way of experiencing danger and developing responses to it.

Music is the one branch of the creative arts which has a close link to science and technology. It has embraced new technologies more enthusiastically than any of the fine arts and used them in a host of ways to improve sound range and quality, to improve the acoustics of concert halls, and to understand the factors that affect the quality of instruments. Unlike activities like sculpture or story-telling, it is not so easy to pin down the origins of human musicality. In '*Musical cheers and the beautiful noise*' I discuss this problem: asking whether we can identify an obvious human activity of a more mundane character that enhanced the survival of those who pursued it, from which musical appreciation or performance can be seen as a by-product. There are a number of quite different possibilities but none is entirely convincing. It may be that a link can be found but it hides well below the surface of music. It appears that human music possesses a similar spectral pattern that is the signature of sound sequences that combine optimal degrees of novelty and predictability. Too much surprise is perplexing whilst too little is boring. These spectral patterns are also the signatures of many natural sound sequences which convey important information in a way that distinguishes them from the random cacophony of background noise. Perhaps here there is a clue as to why our sensitivity to musical sound focuses so much upon these special patterns.

One can go a little further now. The wonderful self-organising property of the sand pile that we saw in Chapter 17 displays the same type of pattern as the musical sequences that we like. The sand pile teaches us that this pattern possesses a property of 'criticality', wherein it is most sensitive to small changes (falls of sand) whilst maintaining a steady overall order (the slope of the pile). In this state small local changes can produce effects over the scale of the entire pile. This is how the overall organisation is maintained. Perhaps our liking for music with this patterning is a reflection of the appeal that criticality has for us. Music that resides in a critical state is going to be the most sensitive to the smallest nuances of timing and performance. This enables it to mean so many different things to different listeners and to provide something new in each individual performance. This is what many music lovers cherish most about musical experience.

The last chapter, '*Stars in their ancient eyes*', takes us to the night sky. We ask how the appearance of the night sky has influenced the development of humanity and the way we think about our position in the Universe. The darkness of the night sky, the odd nature of our solar system that gives rise to eclipses of the Sun and Moon, the sight of the Sun and the blue of the sky: where do they come from? We see something of their inevitability and what alien observers of the Universe might share of our experience of the heavens.

CHAPTER 23

Survival of the artiest

I link therefore I am

S.J. Singer

Cooking is a subject to savour: an art form that stimulates our senses of taste, smell and sight. But the culinary arts had far more menial beginnings. They were provoked by the discovery of fire more than one hundred thousand years ago. The use of fire by humans is universal. Of all living things, only humans cook their food and take trouble to make it look attractive in the process. Cooking makes food easier to consume and digest, kills harmful bacteria and enables food to be preserved for longer, so reducing the time spent gathering it. The simple application of fire to food, while falling well short of the requirements to enter the *Good food guide*, increased the range of palatable foods and rewarded discriminating palates with better health. Thus cookers of food have a clear advantage over eaters of raw meat. In the long run, they will prosper and multiply, thereby passing on their penchant for cooked food to more ancestors than will the non-cookers. From this simple life-supporting activity all manner of exotic dishes have emerged, from crêpe suzette to Yorkshire pudding. The art of culinary expression is a by-product of practices that enhanced the chances of our ancestors' survival in conditions that have long since disappeared.

This style of explanation is one we can use to explore the possible roots of other artistic activities. Is there a core activity that enhances the chance of survival in the long run for those who adopt it? If so, those who adopt it will prosper and multiply at the expense of those who don't. Such practices are called 'adaptive'. They became engendered over millions of years of prehistory, shaping our senses and sensibilities to form the background upon which conscious experience embroiders its own elaborate patterns. These default instincts can influence our aesthetic inclinations. A tantalising case is that of landscape art. Our distant ancestors spent millions of years in tropical savannah habitats —grassy plains with sparsely distributed trees, as we see in parts of Africa today. So, when we now admire pictures of such landscapes, is it because they possess features that once enhanced our chances of survival? Instinctive aesthetic reactions to the world could not have evolved and persisted if, on the average, they lessened the chances of survival. By contrast, responses that increased the chances of survival should, on average, persist. This is why rotten meat tastes unpleasant to us, while sweet citrus fruits taste good. What our hunter-gatherer ancestors did was respond to indicators of safety and survival in a new environment. The appearance of clouds on the horizon is a welcome sight in a dusty savannah grassland, strongly correlated with coming rain and a local abundance of food. A disposition towards finding cloud patterns pleasant would remain as an inherited adaption, which once had positive survival advantages. The savannah landscape displays many faithful cues for safe and fruitful human habitation. These cues are recreated in present-day parklands and recreation areas. There is scattered tree cover that offers shade and shelter from danger, yet there are long vistas with fluent undulations that allow good views and way-finding. Here, too, most food sources are close to the ground. In a forest, by contrast, prey remains out of reach, high above the ground in the tree canopy, and unseen dangers lurk around every dark corner.

There is a clear adaptive advantage to be gained by choosing environments that combine places of security with clear unimpeded views of the terrain—to see without being seen. Many of the scenes we like in artistic representation or architectural construction now reveal a common thread. Expansive views and cosy inglenooks, daunting castles, the treehouse, the *Little house on the prairie*, the mysterious door in the wall of the secret garden: so many classically seductive landscape scenes combine symbols of safety with the prospect of uninterrupted panoramic views or the enticement to explore, tempered by verdant pastures and water. Comfortable pastoral scenes appeal to our instinctive sensibilities because of the selective advantages that such environments first held for our ancient forebears. Their style figures prominently in our best-appreciated landscape gardening, public parks and gardens, where they relieve stress and encourage feelings of well-being. Architects often use sloping ceilings, overhangs, gabling and porches to create the feeling of refuge from the outside world; while balconies, bays and picture windows meet our desire for a wide-ranging view that allows us to see without being seen.

We find flowers beautiful, therapeutic and romantic. What still-life subject is more common? Our unusual interest in colourful flowers and the lengths we go to to cultivate and arrange them are impressive. We don't eat flowers but they give information about the ripeness or rottenness of fruit and their appearance is a useful cue to rapid identification of different plant-forms. If no flowers are present, all plants look green and can be distinguished only by detailed inspection. Sensitivity to flowers thus has a purpose, which is adaptive, and provides us with a clue to the origin of what would otherwise be an entirely mysterious human fascination.

Flowers, like all living things, possess obvious symmetry. All through the visual arts we find that symmetry entices the eye and the brain. We delight in the intricacies of a drawing by Maurits Escher or the fabulous tilings of the Alhambra.

The acuteness of our visual ability to detect symmetry might well have origins in the simple fact that almost all living things possess lateral (left–right) body symmetry while inanimate objects do not. Our evaluation of human physical beauty focuses on the symmetry of facial and bodily form; cosmetic surgeons make very large sums of money restoring or enhancing it. Sensitivity to symmetrical patterns in a confused or camouflaged scene is a life-saving sense that helps us pick out potential predators, prey or mates. Indeed, one can appreciate why it might pay to be a little oversensitive to patterns in this respect: better to be thought paranoid once in a while because you saw tigers in the bushes when they weren't there than to end up as someone else's lunch because you didn't see them when they were there. Sensitivity to pattern and symmetry is adaptive.

It has become fashionable to regard human aesthetic preferences as totally subjective responses to learning and nurture. This now seems barely credible. Our sensitivities and emotional responses have not sprung from nothing. The evaluation of environments was a crucial instinct for our distant ancestors—one upon which their survival, and hence our existence, hinged. The adaptive responses that we have inherited over millions of years form a base over which subsequent social experience is laid. In many visual arts, we see remnants of past imperatives, now invested with symbolism or subverted into opposition, perhaps, but undeniably present in our representations of natural landscapes. Even amid artificial symbolisms of a religious or romantic sort, one can often find a background resonance with echoes of our innate emotions.

Finally, we should take up the challenge of music. While there have been cultures without counting, cultures without farming, cultures that made no use of science or writing, we know of none without music. Such universality provokes us to ask whether it too is a by-product of a much simpler ancient activity that increased the chances of survival. Charles Darwin thought that music-making emerged from mating calls like birdsong. Perhaps it is an outgrowth of using sound to send signals or to imitate animal calls when hunting. A deeper clue may lie in its emotive power. Wherever we find a need to reinforce group solidarity or inspire acts of bravery, we find music. It creates an atmosphere within which ideas and signals can register a strong impression upon the human mind. It both calms and arouses. Particularly impressive is the effect of rhythmic drumming because we feel as well as hear it. Percussion is the most basic of sounds. It is always present in ancient ceremonies of initiation or religious worship. It binds together the thoughts and actions of individuals to produce a shared experience that makes them stronger and bolder than they might otherwise be.

Another avenue is to examine the pattern of sounds within music. The most intriguing discovery has been that the musical sound patterns appreciated across a wide range of human cultures display the optimal balance between predictability and surprise. This pattern is also found in many natural noise sources and our hearing and sound sensitivity may have evolved to be most acutely sensitive to this underlying pattern because it carries the most vital information about the environment. The more sensitive we are to it in the face of other random sounds, the more likely we are to avoid danger. With the passage of time and the complex elaborations of consciousness, this conditioning has given rise, as a by-product, to our gifts of appreciation. Why music affords pleasure is because it indulges our liking for such patterns to the full.

CHAPTER 24

Musical cheers and the beautiful noise

It is quite untrue that English people don't appreciate music. They may not understand it but they absolutely love the noise it makes.

Thomas Beecham

There have been cultures without counting, cultures without painting, cultures without science, cultures bereft of the wheel and the written word, but never a culture without music. Music, scented sound, is there in the jungle, in the city, between our ears and at our fingertips. Without consciously learning its rules we can respond to its rhythms. Musical ability among the very young, like mathematical genius, can be alarmingly sophisticated—out of all proportion to other skills. Some people find that music is a necessary accompaniment for the successful completion of other activities. But its definition is not so easy because music covers a vast range of sound levels and frequencies—from simple repetitive drumming to symphonic works of enormous complexity in which the mental powers and dexterity of many individuals combine to represent the patterns encoded in the score.

When one finds transcultural human activities, like writing, speaking, and counting, that display common features, we can ask if they have evolved from simpler activities whose *raison d'être* is more obvious. If the simpler predecessor of today's complex activity endowed its exponents with a clear advantage in life, because it made them safer, healthier, or just plain happier, then it is likely to become more prevalent because of its cultural transmission or, if it derives from some inheritable genetic trait, by becoming more likely to survive and be inherited.

At first sight it is not easy to see what advantage a penchant for Beethoven or the Beatles confers on its possessor. What could have been the utility of such an abstract form of sound generation and appreciation? This is a difficult riddle. Musical appreciation might be merely a by-product of mental attributes evolved for quite different purposes. There are plenty of possibilities to choose from. The earliest, most spontaneous of human sounds is the cry of a baby at birth, when hungry, or when distressed. Sounds that are responded to in circumstances of great intimacy and emotion. But humans of all ages retain the ability to make similar sounds and emotional cries and there is no similarity between those cries and music. Indeed, we recognise the instinctive reaction to crying to be one of irritation, unease, or distress—just the reaction we might have expected experience to have reinforced upon our ancestors, but not our response to most forms of music. A clue to music's antecedents may lie in its emotive power. In civilizations ancient and modern, the world over, we find the sound of music whenever there is a need to increase group bonding or inspire acts of courage. But herein lies a paradox. For we find that music calms the overwrought human mind as effectively as it can rouse it. This dichotomy suggests that we will not find the source of musical performance or appreciation in so peculiar a function as the arousal or subduing of specific human emotions. Nor, despite an ancient tradition going back to Pythagoras, does there seem to be a deep connection between mathematics and music. Mathematics is the study of all possible patterns; but music is more than patterns in sound. It is resonant with something. Music can stir mass emotion, inspire nationalism, and religious fervour. Mathematics can do none of these things. The 'music of the spheres' is silent and abstract.

An ubiquitous source of sound is the inanimate natural world: the wind, the rush of running water, the crash of thunder. Alas, these are things that one seeks to be heard in spite of; they hardly qualify as templates for human emulation unless you were attempting to camouflage your presence while hunting or hiding. More promising models are sounds from elsewhere in the living world. Mating calls and complex bird-song play a well-defined role in the evolutionary process: mates are attracted and territory demarcated. Darwin thought that music had its prehuman origins in mating calls.

The most impressive feature of music is its temporal continuity. Whereas art displays pattern in space, music offers patterns in time. Just as the mind has developed acute pattern recognition abilities, so it possesses exquisite sensitivity to nuances of sound over a far greater range than it does for visual images. This is part of a wider facility. We have developed ways to make sense of time in ways that transform chains of events into a history. Legends and traditions first played this role and complemented the human understanding of events. The spatial order exhibited in painting or sculpture is heightened when endowed with a temporal aspect. This is why films are often more appealing than still photographs and why children can find video games so addictive. Unchanging images leave the viewer to look for themselves. They can look again and again, first following one sequence, then another. But music has its own sequential order of perception. It has a beginning and an end. A painting does not. Thus we see that music may be associated with a need to structure time or be derived as a by-product of an advantageous adaption for a structuring of time. What sort of advantage could this offer?

Sequential 'timing' is something that lies at the heart of all manner of complex human activities, from juggling to starting a car, which require meticulous coordination of eye, brain, and hand. Take crossing the road: we receive visual and sonic information about vehicles moving in different directions at different unknown speeds, viewed at differing angles under variable light; we need to evaluate whether there is an interval of time sufficient to cross the road and then move appropriately. Viewed in this light, it seems astonishing that we ever manage to make it from one side of the street to the other. The brain has clearly developed an extraordinary facility for sequential and parallel timing of different movements that will combine to produce a rapid, single continuous activity like the serving of a tennis ball. Music could be a by-product of the complicated mental circuits that evolved primarily in response to the adaptive advantage offered by an ability to coordinate body movements in precise, continuous, and rapid response to outside changes. The obvious limitation of this type of explanation is the absence of musical inclination in those apes who are the planet's unchallenged gymnastic champions.

If we have evolved to cope with the changing patterns of a complex environment, there may be naturally occurring forms of complexity which our brains are best adapted to apprehend. Thus, artistic appreciation might emerge as a by-product of those adaptations. An interesting aspect of musical sound has recently come to light and suggests how this might occur.

Physicists and engineers prosaically refer to sequences of sound as 'noises'. A useful way to distinguish them is by their spectrum: this is a measure of the distribution of intensity over wave frequencies (just as the spectrum of light revealed by a prism displays the distribution of light with colour, which is just another word for frequency in this context). An important feature of many noise spectra is that their intensities are proportional to a mathematical power of the sound frequency over a very wide range of frequencies. In this case, there is no special frequency that characterises the process—as would result from repeatedly playing the note with the frequency of middle C. Such processes are called scale-free. In a scale-free process, whatever happens in one frequency range, happens in all frequency ranges. Scale-free noises have intensity spectra that are proportional to inverse powers of the frequency, $1/f^a$, where a is a constant number. Their character changes significantly as the value of a changes. If noise is entirely random, so that every sound is completely independent of its predecessors, then a is zero, and the process is called 'white noise'. Like the spectral mixture that we call white light, white noise is acoustically 'colourless' —equally anonymous, featureless, and unpredictable at all frequencies, and hence at whatever speed it is played. When your television picture goes haywire, the 'snow' that blitzes the screen is a visual display of white noise produced by the random motion of the electrons in the circuitry. At low intensities, white noise has a soothing effect because of its lack of discernible correlations. Consequently, white noise machines are marketed to produce restful background 'noise' that resembles the sound of gently breaking ocean waves. White noise is invariably 'surprising', in the sense that the next sound cannot be anticipated from its predecessor. By contrast, a scale-free noise with $a = 2$ produces a far more correlated sequence of sounds, called 'brown noise'. This is also unenticing to the ear; its high degree of correlation gives it a predictable development, like a musical scale. It 'remembers' something of its history. Brown noises leave no expectation unfulfilled, while white noises are devoid of any expectations that need to be fulfilled. But midway between white and brown noise, when $a = 1$, lies the special case of '$1/f$ noise'. It is special because such signals are moderately correlated, and hence possess 'interesting' patterns over all time intervals. They combine novelty with expectation in an optimal way.

In 1975, Richard Voss and John Clarke, two physicists at the University of California at Berkeley, serendipitously discovered that many classical and modern musical compositions are closely approximated by $1/f$ noise over a very wide range of frequencies. Appealing music exhibits an optimal level of novelty at the spectral level: neither too predictable like brown noise, nor randomly unpredictable like white noise. This may be telling us important things about the mind's first adaptations to the world of sound. But if musical appreciation is a by-product of a more general pattern-processing propensity of the brain, why are our senses heightened by $1/f$ noises? It is significant that the world around us is full of variations with $1/f$ spectra. One reason is the prevalence of sequential processes in the natural world. Benoit Mandelbrot has claimed that our nervous system acts as a spectral filter, preferentially passing $1/f$ noise to the brain while filtering out white noise at the periphery to prevent the brain being swamped with uninteresting random background noise about the world. An optimal response to signals with $1/f$ form might well be the best investment of resources that the system can make for information gathering, or it may be the simplest way for the nervous system to decode vibrations in the inner ear.

Music is the purest art form. Our minds receive a sequence of sounds woven into a pattern—largely undistracted by the other senses. The fact that a wide range of music exhibits a $1/f$ spectrum for its variations in loudness, pitch, and interval, across the whole range from classical to jazz and rock, suggests that this appeal arises from an affinity for the statistical features of natural noises, whose detection and assimilation were adaptively advantageous to humans. This affinity extends to many varieties of non-Western music; there is a good approximation to $1/f$ noise in all of the musical traditions studied: from the music of the Ba-Benzele Pygmies, traditional Japanese melodies, Indian ragas, Russian folk music, to American blues. It would be particularly interesting if a traditional musical culture (or a whalesong) could be found where there was a significant, habitual deviation from the $1/f$ spectral form.

One should not regard this argument as totally reductionist, any more than one should take seriously music-lovers' claims that music is a transcendental form with charm beyond words. Our minds, with their propensity to analyse, distinguish, and respond to sounds of certain sorts, yet ignore others, have histories. Musicality seems most reasonably explained as an elaboration of abilities that were evolved originally for other more mundane, but essential, purposes. Our aptitude for sound processing converged upon a sensitivity for certain sound patterns, because their recognition optimised the reception of vital information. With the emergence of that more elaborate processing ability we call consciousness, has arisen an ability to explore and exploit our innate sensibility to sound. This has led to organised sound forms that span the range of pitches and intensities to which the human ear is sensitive. Those forms diverge in their stylistic nuances, as do the decorations in peoples homes, from culture to culture. But the universality of musical appreciation, and the common spectral character of so much of the sound that we embrace, behoves us to look at the universal aspects of early experience for an explanation. Had the sounds that fill our world been different in their spectral properties, we would have developed a penchant for sounds with quite different structures—structures that from the spectral perspective would have been more surprising or more predictable.

Alternatively, and more likely, is that the source of $1/f$ noise in our environment is a consequence of statistical processes of such generality that this form of noise would be ubiquitous in any environment—terrestrial or extraterrestrial. In that case, we might have been overly pessimistic in believing that music could not be used to communicate with extraterrestrials. If, as we might expect, their environments display a cacophony of vital $1/f$ spectral variations, they should have evolved a special sensitivity to them. When transformed into the appropriate medium, they might well appreciate some of our music—which is just as well because the *Voyager* spacecraft, now heading out of the solar system to the stars, contains an elaborate recording of terrestrial sounds. Ninety minutes of music was included—Bach, Beethoven, rock and jazz, together with folk music from a variety of countries. The senders did not know it, but it all has a $1/f$ spectrum.

CHAPTER 25

Stars in their ancient eyes

I have loved the stars too fondly to be fearful of the night.

Sarah Williams

The Christmas story follows a star—or at least a triple conjunction of Saturn and Jupiter in Pisces. Yet, surrounded by the nocturnal glare of artificial light that emanates from our cities, we see little of the stars. They have no significance at all for most of us. For the ancients, especially those living below clear skies or in the rarefied air of mountainous regions, things were very different. The spectacle of the star-spangled sky would have been the most impressive sight of their lives. No wonder that stories grew up in which these pieces of celestial jewellery played a starring role.

At the beginning of this century, the philosopher George Santayana picked out the appearance of the night sky as an exemplar of what is attractive to the human mind: a level of intricacy delicately poised between bewildering complexity and banal simplicity. A hint of intriguing pattern challenges the mind to ponder and seek it out. If we were to see the night sky for the first time, imagine the consequences of such an astral awakening. It was just such a contemplation that inspired the young Isaac Asimov to pen his famous short story 'Nightfall' about the final days of the civilisation on the planet Saro. This strange world basked in the light of six suns. Natural darkness was unknown; so, therefore, were the stars. The inhabitants evolved in a world of light with a strong fear of the dark, and an understandable susceptibility to claustrophobia. Their astronomers were convinced of the smallness of the Universe. Unable to see beyond their own sixfold solar system, they contented themselves with showing how its complicated motions could be understood using the same law of gravitation that worked so well on the surface of Saro. These rationalists shared their world with romantic cultists, who perpetuated an 'old knowledge' of a world of stars beyond the sky and a coming of darkness when the world would end. Social tensions in Saro mount when the astronomers discover that there must be an unseen moon in their solar system that will only become visible when it eclipses one of the suns. A few astronomers realise that the moon will eclipse the second sun of the system, at a time when it is the only sun in the sky. The eclipse will be total. News of this expectation leaks out. Civil unrest mounts as the cultists stir up eschatological fever. Darkness blots out the sky and tens of thousands of brilliant stars appear, shrouding the planet in a canopy of twinkling starlight. For Saro is not denizen of the sparsely populated stellar suburbs of a galaxy like the Milky Way; it lies deep in the dense heart of a star cluster. Panic and civil unrest breaks out. There, the story ends; the reader is left to ponder the revolution in outlook that is about to occur.

The simple question of why the sky is dark at night is surprisingly subtle and nothing to do with the Sun. It was first raised by Edmund Halley (he of comet fame) in 1721, before being resurrected by Wilhelm Olbers, a German astronomer, in 1823. Nowadays, it is known as Olbers' paradox. A paradox because it appeared that a Universe filled with stars should result in our line of sight ending on a star in whatever direction we look. When we look into a dense forest our line of sight always ends on a tree, so when we gaze out into the Universe why does the whole night sky not look like the surface of an overlapping forest of stars?

The simple answer to Olbers' paradox seems to be that there is too little matter in our Universe to produce enough energy to make the entire sky bright. If you could snap your fingers and convert all the material in the Universe into radiation it would merely raise the temperature of the ambient cosmic background radiation from three to 30 degrees above absolute zero. We would not notice anything amiss. But the reason the Universe contains such a low density of matter is because it has been expanding and rarefying for more than ten billion years. This vast expanse of time is not an example of cosmic procrastination. It is needed for stars to form and then produce the elements heavier than helium that are required for the spontaneous evolution of any form of chemical complexity. The dark night sky is a by-product of those conditions that are required if a universe is to be habitable by living things. Once, billions of years ago, when the universe first emerged from the Big Bang, the sky was bright—brighter than a billion Suns—but no one was there to see it.

The influence of the night sky upon our own civilisation has not been quite so cataclysmic as it was on the fictional world of Saro, but it has nevertheless been deep and far reaching. By coincidence the respective distances from us of the Sun and the Moon conspire to present the same apparent sizes in the sky, despite the huge difference in their true sizes. As a result we see complete eclipses of the sun.

Ancient eclipses are famous for their influence upon human affairs. The total eclipse that occurred on May 28 in 585 BC was so dramatic that it ended the five-year-old war between the Lydians and the Medes. Their records tell us that in the midst of battle 'the day was turned into night'; at once their fighting stopped and a peace treaty was signed. In contrast, the eclipse of the Moon on August 27 413 BC brought about a less amicable end to the Peloponnesian war between the Athenians and the Syracusians. The Athenian soldiers were so terrified by the eclipse that they became reluctant to leave Syracuse, as planned. Interpreting the eclipse as a bad omen, their commanders delayed the departure for a month. This delay delivered all their forces into the hands of the Syracusians—the procrastinating commander was put to death. In 1503 Christopher Columbus exploited his knowledge of an eclipse of the Moon by the Sun to enlist the help of the Jamaicans after his damaged ships were stranded near their island. At first, he traded trinkets with the natives in return for food, but after a while they refused to provide any more provisions. Columbus's response was to arrange a conference with the natives on the night of February 29, 1504—the time when an eclipse of the Moon would begin. He announced that his God was displeased by their lack of assistance and was going to remove the Moon as a sign of his deep displeasure. As the Earth's shadow began to fall across the Moon's face, the natives quickly agreed to provide him with anything he wanted, so long as he brought back the Moon. Columbus informed them that he would need to withdraw and persuade his God to restore the lesser light to the heavens. After retiring with his hourglass for an appropriate period, he then returned in the nick of time, to announce the Almighty's pardon for their sins and the restoration of the Moon to the sky. Soon afterwards, the eclipse ended. Columbus had no further problems on Jamaica; he and his men were subsequently rescued, and returned in triumph to Spain.

Thus, there is more to the sky than meets the eye. Our time-keeping is a curious mixture of astronomy and astrology. The days are dictated by the period of the spinning Earth, the months by the Moon's motion, the year by our long slow orbit of the Sun, and the seasons by the inclination of the Earth's polar axis to the plane of that orbit, and the week by the number of large bodies periodically wandering through the heavens in ancient times. Our uncertain sense of place in the Universe and our feelings for the transcendental owe much to the vastness and darkness of space around us. Our mythologies and legends are infused with personifications of the sky's patterns, our science rests upon the insights we gained by the precise measurement of its clockwork. Most striking of all, the rays of sunlight that fall upon the Earth's atmosphere create the clear blue sky, the red of sunset, the clouds of white that fill the world with colour. Sunlight makes colour vision advantageous and periodically promotes so much of that extravaganza of diversity that we call life. But when the darkness comes the whole world changes again.

Part 7

Time

Time is a spring that flows from the future
Miguel de Unamuno

Time has taken a curious grip on the human imagination. Is it that we are frightened by a lack of it, or hard-pressed to find something to do with it? Stephen Hawking's amazingly successful 'brief history' of our attempts to understand the relationship between time and the Universe launched a flotilla of imitations and stimulated physicists and philosophers to probe more deeply into the problem of distinguishing the past from the future. This section begins with an account of the way that modern cosmologists think about time in the Universe and what happens when one tries to marry that picture to the uncertainties of quantum mechanics. Curiously, the choices that modern science offers us as to the true nature of time include those that would have been familiar to St Augustine more than 1600 years ago. But the work of Jim Hartle and Stephen Hawking has produced the new possibility that our journey into the past to find the beginning of time leads not to a first moment but a gradual melting away of the notion of time into quantum fuzziness. Perhaps time is not truly fundamental but merely another of those things that characterise universes that are big and old, dark and cold, and far from the extremes of density and temperature in which the quantum theory reveals the true colour of its money. This first chapter offers a systematic account of these developments in cosmological thinking.

A more informal piece, '*Universe began in no time at all*', encapsulates some of the nagging problems and possibilities concerning the nature of time. Like 'chaos', time has become fashionable again amongst novelists and film-makers, witness Martin Amis' *Time's arrow*, Alan Lightman's delightful stories, *Einstein's dreams*, about worlds where time is different, and a whole raft of movies and TV series that use the central device of time travel around which to spin their plots. Perhaps our fascination springs from the asymmetry in the way that time and space challenge our existence. We have been much more successful in overcoming the limitations of space than those of time. We have greatly expanded the domain of life, filling most of the Earth with people, and we have even begun to survey the solar system for sites where outposts of life might be planted. But our battle with time has been less successful. We may have doubled human life expectancy in the last 200 years, but death is as certain as ever. The limits of space do not yet threaten our continued existence like those of time.

The last chapter, entitled '*Arrows and some banana skins*', is a review of a book which attempts a very careful analysis of the problem of the asymmetry between the past and the future that we find all over the place in our experience. Our subjective experience of time distinguishes clearly between the past and the future; our astronomical experience shows that the Universe is expanding towards the future; the second law of thermodynamics is the physicists' way of distinguishing the past from the future since disorder appears to increase in closed systems as time passes. But are these different 'arrows' of time connected?

Some cosmologists have been keen to apply the notion of entropy's inexorable increase to the Universe as a whole. If one could identify the entropy of the Universe and show that it is very small today then the requirement that it be always increasing would tell us that the Universe must have begun in a fantastically well-ordered state because it is so old. Unfortunately, this argument does not quite work. We have not been able to discover what the entropy of even the observable part of the Universe is, never mind that of the entire Universe. We have isolated some of the contributions to it but it remains to be seen what we might have left out. The entropy of the Universe may be bigger than we think and that of the portion that we can see may be very different from that beyond our horizon.

Throughout these discussions the reader will detect an interesting contrast between two views of time. On the one hand there is the view of the astronomer and the cosmologist, who see the Universe as a great chunk of space and time. This chunk can be sliced up, like a piece of cheese, in innumerable ways. Each slicing defines another way of keeping time. None is fundamentally superior to any other. Each observer's 'slicing' of space–time into space and time is determined by their motion. The fact that there are many ways of defining time is just the 'relativity' of time that Einstein discovered. A curiosity of this perspective is that it seems to require the future to be already 'out there' waiting for us to enter. For someone using a different slicing, events in our future lie in their present. There is no absolute meaning to the concept of 'now' or of different events being 'simultaneous'. In complete contrast, is the perspective on time that is provided by sciences like chemistry, where the impact of the second law of thermodynamics can be very impressive. The passage of time is clearly evident in the events that we witness locally. The trend from order to less order distinguishes the past from the future in a way that seems more local, more mundane, and less fatalistic than the picture of the cosmologist. Whose viewpoint is more fundamental? The betting is on the cosmological viewpoint, but the smart money will stay in punters' pockets for a while. It is clear that there is still much to learn about the nature of time. We are only scratching the surface of the cosmological picture of time in the presence of quantum uncertainty. The mysteries of time's disappearance in the distant past may seem fantastic but if this idea is mistaken I would bet that what supersedes it is more fantastic still.

CHAPTER 26

Cosmological time

Time is at the heart of all that is important to human beings.

Bernard d'Espagnet

There is a long-standing philosophical puzzle regarding the Nature of time that has emerged in the works of different thinkers over the millennia. It reduces to the question of whether time is an absolute background stage on which events are played out but yet remains unaffected by them, or whether it is a secondary concept wholly derivable from physical processes and hence affected by them. If the former picture were adopted then we could talk about the creation of the physical Universe of matter *in time*. It would be meaningful to discuss what occurred before the creation of the material Universe and what might happen after it passed away, Here, time is a transcendent part of reality without a conceivable beginning or end. This idea lends itself readily to the Platonic notion that there exist certain eternal truths or blueprints from which the temporal realities derive their qualities. Indeed, time takes upon itself many of the qualities traditionally associated with a deity. The alternative, an idea that emerges in Aristotle's writings and in those of the early Islamic natural philosophers, before being reiterated most memorably by St Augustine and Philo of Alexandria, is that time is something that comes into being with the Universe. Before the Universe was, there was no time, no concept of before. Such a device enabled the medieval Scholastics to evade difficult conundrums about what took place before the creation of the world and what the deity was doing in that period. In essence this views time as a derived phenomenon, inextricably bound up with the contents of the Universe. The beginning of time is the moment when constants and laws of Nature must come into being ready-made and ready to go. In *The city of God* St Augustine writes, 'then assuredly the world was made, not in time, but simultaneously with time. For that which is made in time is made both after and before some time—after that which is past, before that which is future. But none could then be past, for there was no creature, by whose movements its duration could be measured. But simultaneously with time the world was made.'

This is close to our common experience of time. We measure time using clocks which are made of matter and which obey laws of Nature. We exploit the existence of periodic motions, whether they be revolutions of the Earth, oscillations of a pendulum, or vibrations of a caesium crystal; and the 'ticks' of these clocks define the passage of time for us. We have no everyday meaning to give to the notion of time aside from the process by which it is measured. We might thus defend an operationalist view wherein time is defined by its mode of measurement alone.

Whereas in the transcendental view of time we might speak of bodies moving *in* time, the emphasis of the latter view is upon time being defined by the motion of things. One of the advantages of the first view is that one knows where one stands and what time is always going to look like—it is the same yesterday, today, and forever. By contrast, the second picture promises to produce novel concepts of time—and might even do away with the concept altogether—as the material contents of the Universe alter their nature under varying conditions. We should be especially conscious of such a possibility as we backtrack towards those moments of extremis in the vicinity of the Big Bang. For any moment that appears to be the beginning of time inevitably exists where the very notion of time itself is likely to be most fragile. In an expanding and constantly changing Universe, the operational view of time is likely to produce a subtle and variable conception of time's status and meaning.

The image of a transcendent, absolute time shadowing the march of events upon a cosmic billiard table of unending and unchanging space was the foundation of Newton's monumental description of the world in which he announces that: 'I do not define time, space, place and motion, as being well known to all. Only I must observe, that the common people conceive those quantities under no other notions but from the relation they bear to sensible objects. And thence arise certain prejudices . . .' Once the equations governing the change of the world in space and time are given then the whole future course of events is determined by the starting conditions. Time appears superfluous. Everything that is going to happen is programmed into the starting state. (This will not be true if other physical processes become involved. For example, in the archetypal situation of billiard balls moving according to Newton's laws, their future behaviour after collisions depends upon the rigidity of the collisions and this involves knowledge of the behaviour of the materials out of which the balls are made. This information is beyond the scope of Newtonian mechanics.)

The Newtonian laws of motion could be applied to the description of the world and followed backwards in time. Our Universe is observed to be expanding and hence a Newtonian description leads to the assertion that there must have been a past moment of time at which everything was compressed to zero size and infinite density—the 'Big Bang' as it was first termed by Fred Hoyle.

However, because of the absolute nature of space and time in the Newtonian world view, we cannot draw any conclusions about the Newtonian Big Bang constituting an origin to time, let alone the origin of the Universe. It is simply a past time at which known laws predict that some physical quantities become unboundedly large; we say they become infinite in value there. But space and time go on regardless.

The first scientists to contemplate the significance of places where things apparently cease to exist or become infinite—'singularities', as we would now call them—in Newtonian theory were the 18th century scientists Leonhard Euler and Roger Boscovich. They both considered the physical consequences of adopting force laws for gravitation other than Newton's famous inverse-square law. They found some of the alternatives had the unpleasant feature that the solutions just cease to exist after some definite time in the future when one studied the behaviour of objects orbiting around a central sun. They cannot be continued forwards any further in a world governed by one of these maverick force laws. Boscovich thinks it absurd that the body must disappear from the Universe at the centre if the force law were inverse-cube rather than inverse-square. He draws attention to Euler's earlier study of motion under the influence of gravity where the master mathematician, 'asserts that the moving body on approaching the centre of forces is annihilated. How much more reasonable would it be to infer that this law of forces is an impossible one?' These appears to be the first contemplations of such matters in the context of Newtonian mechanics.

In fact there are deep problems with attempting to apply Newton's theory of gravity and motion to the Universe as a whole. If we try to make space finite in volume then the density of matter is forced to be zero. (Newtonian universes can be infinite in volume and contain only a finite quantity of mass.) The same is true if one tries to make a Newtonian universe finite in volume by curving up its space or joining opposite sides to make its topology finite. Thus, the Newtonian world requires the Universe of matter to be a finite island of matter in an ocean of infinite absolute space or an infinite island of matter.

Worse still, Newton's theory is incomplete. It does not contain enough equations to tell us how all the allowed changes to the Universe actually occur. If the Universe expands or contracts at exactly the same rate in every direction then everything is indeed determined, but when any deviations from perfectly spherical expansion are allowed at the start then determinism breaks down, for there are no Newtonian laws that dictate how the shape of the world will change with time. Clearly, Newton's theory of absolute space and time is defective. The next step to take is to contemplate some coupling of the notions of space and time to the material contents of the world.

The earliest and most intriguing speculation of this sort was made by

William Clifford, an English mathematician who contemplated just the type of situation that Einstein would build into the general theory of relativity. Clifford was motivated by the mathematical investigations of Riemann who had formalised the geometric study of curved surfaces and spaces that possess non-Euclidean geometry (that is, the three interior angles of a triangle no longer add up to 180 degrees, where the three corners of the triangle are formed by joining the shortest lines that can be drawn between them to form the sides of the triangle on the curved surface). Clifford appreciated that the traditional space of Euclid is thus one of many and we can no longer assume that the geometry of the real world possesses the simple Euclidean form. The fact that it appears to be flat locally is not persuasive because most curved surfaces appear flat when viewed over small areas. After studying Riemann's ideas Clifford proposed this radical scenario in his paper of 1876. In 1900, the German astronomer Karl Schwarzschild applied these ideas to the Universe for the first time. He proposed that space might be curved and used star counts to place limits on the possible curvature and size of space.

This prescience is rather remarkable. Although Einstein never seems to have been aware of these remarks, Clifford's intuitive idea became the central idea of the general theory of relativity. The geometry of space and the rate of flow of time are no longer absolutely fixed and independent of the material content of space and time. The matter content and its motion determine the geometry and the rate of flow of time, and symbiotically this geometry dictates how matter is to move. Einstein's elegant theory of gravitation possesses a set of equations that dictate the connection between the matter content of the Universe and its space and time geometry. These are called field equations and they generalise the Newtonian field equation of Poisson, which encapsulates Newton's inverse-square law of gravitation. In addition to this structure, there exist equations of motion that give the analogues of straight lines in the curved geometry. These generalise Newton's laws of motion.

One further erosion of time's absolute Newtonian status occurs in Einstein's theory. Einstein's theory was built upon a premise that there are no preferred observers in the Universe: no set of observers for whom all the laws of Nature look simpler. The laws of physics must have the same form for all observers no matter what their state of motion. That is, however your laboratory is moving—whether it is accelerating or rotating with respect to that of your neighbour—you should both find the same laws of physics to hold. You may each measure observables to have different values but you will, none the less, find them to be linked by the same invariant relationships.

In Einstein's world there is no special class of observers for whom, by virtue of their motion and time-keeping arrangements, the laws of Nature look especially simple. This is not true in Newton's formulation of motion. His famous laws of motion are found to hold only by experimenters moving in laboratories that are in uniform, non-rotating motion with respect to each other and with respect to the most distant stars, which he took to establish a state of absolute rest. Other observers who rotate or accelerate in unusual ways will observe the laws of motion to have a different, more complicated form. In particular, and in violation to Newton's famous first law of motion, they will observe bodies acted upon by no forces to accelerate.

This democracy of observers that Einstein built into the formulation of his general theory of relativity means that there is no preferred cosmic time. Whereas in his special theory of relativity there could exist no absolute standard of time: all time measurements are made relative to the state of motion of the observer; in the general theory of relativity things are different. There are many absolute times in general relativity. In fact, there appears to be an infinite number of possible candidates. For instance, observers around the Universe could use the local mean density or expansion rate of the Universe to coordinate their time-keeping. Unfortunately, none of these absolute times has yet been found to possess a more fundamental status than the others.

A good way to view an entire Universe of space and time—a 'space–time'—in Einstein's theory is as a stack of spaces (imagine there to be only two dimensions of space rather than three for the sake of visualisation), with each slice in the stack representing the whole Universe of space at a different time. The time is a label identifying each slice in the stack. The discussion of the previous paragraph means that we can actually slice up the whole space–time block into a stack of 'time slices' in many different ways. We could slice through the solid stack at a variety of different angles. This is why it is always more appropriate to talk about space–time rather than the somewhat ambiguous partners, space and time. But the connection between matter and space–time geometry means that 'time' can be defined *internally* by some geometrical property, like the curvature, of each slice and hence in terms of the gravitational field of the matter on the slice that has distorted it from flatness (see Figure 26.1 for a simple illustration). Thus we begin to see a glimmer of a possibility of associating time, including its beginning and its end, with some property of the contents of the Universe of the laws that govern how they change.

The new picture of *space–time* rather than space *and* time considerably changes our attitude towards initial conditions and the possible beginning of the Universe. Because of the coupling that exists between the fabric of space–time and matter, any singularity in the material content of space–time—for example, the infinity in the density of matter which occurs in the traditional picture of the Big Bang—signals that space–time has come to an end as well. We now have singularities *of* space and time not merely singularities *in* space and time. Moreover, any space–time given by Einstein's theory of general

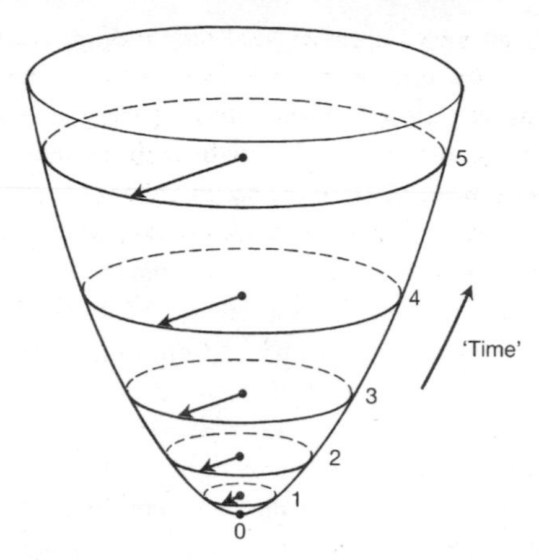

Fig. 26.1. Each of the slices labelled 1,2,3, and 4 taken through space can be endowed with a 'time' label that is determined by the radius of the arrowed circle. As we progress up the curved paraboloidal surface, the increase of 'time' is recorded by the increase in the radius of the circle bounding the slice.

relativity is an entire Universe. Unlike in Newton's theory, it can never merely describe some object sitting on an external stage of fixed space. Thus, the singularities of general relativity are features of the entire Universe, not just one place in it or one moment of its history. These singularities mark out the boundary of space and time.

If we study the expanding Universe according to this picture and trace its history backwards then it is possible for it to begin at such a singularity. This prediction has been seized upon by many as proof that the Universe had a beginning in time. However, like any logical deduction, this conclusion follows from certain assumptions whose truth needs to be closely examined. The most shaky of these assumptions is that gravity is always attractive. Our modern theories of elementary particles contain many types of particle, and forms of matter, for which this assumption is not true. Indeed, the whole inflationary universe picture is founded upon the requirement that it *not* be true; for only then can the brief period of accelerated 'inflationary' expansion arise. However, although the avoidance of a singularity might avoid a beginning to time it would not save us from having to prescribe 'initial' conditions at some past moment to select our actual Universe from the infinity of other possible worlds that begin at singularities. Even if there did exist a singularity, one must face the fact that there are different types of singularity. The specification of the properties of this singularity is an 'initial' condition to be specified on the boundary of our space and time. Some extra ingredient still needs to be found that could provide that specification.

General relativity (and any other relativistic theory of gravity which does not possess absolutely fixed space or time) gives rise to another subtle property not present in simple Newtonian conceptions of space and time. There are actually many distinct space–times that can arise from the same initial conditions.

Suppose that some space–time, S, has initial conditions set at some starting time zero, which we shall label $t(0)$. We can construct another space–time by removing all of that part of the first space–time that lies to the future of some time $t(1)$ later than $t(0)$ as well as the time $t(1)$ itself. The new space–time, S', is the same as S to the past of the moment $t(1)$ but contains no space or time whatsoever to the future of $t(1)$, as illustrated in Figure 26.2.

But both S and S' arise from the same initial state and, indeed, we could have cut pieces off S in an infinite number of different ways to make other space–times that start from the same initial conditions. Yet there is something unsavoury about S' and its fellow neutered universes. It comes to an end at the allotted time, $t(1)$, for no physical reason whatsoever. There is no singularity of any physical quantity. Indeed, we have not even had to make mention of the material contents of the Universe at all. The equations that govern the behaviour of matter would like to predict the future if only you would allow there to be a future.

This arbitrary truncation of the future is regarded as unrealistically artificial and cosmologists choose to exclude its possibility and specify the future

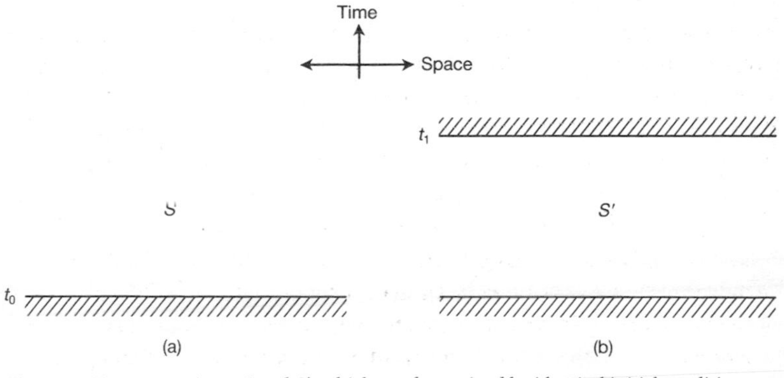

Fig. 26.2. Two space–times, S and S', which are determined by identical initial conditions on the initial time slice t_0. In (a) the space–time S is *maximally extended*, whereas in (b) it is terminated at some time t_1, but no physical infinity or other barrier to its future continuation arises then. The space–time S' is therefore identical to S until the time t_1, but does not exist to the future of t_1. In practice, cosmologists always assume that a given set of initial conditions physically realises the maximally extended space–time and not one of the infinite number of alternatives that are identical to it until some arbitrary moment when they cease to exist.

evolution uniquely. To do so, it is necessary to introduce a further condition into the prescription of possible space–times, or universes, in theories like general relativity, in addition to the specification of initial conditions and laws of Nature. One requires that the Universe should continue to exist until the laws of Nature governing the behaviour of mass and energy signal that time itself has come to an end at a real physical singularity. Under reasonable conditions it transpires that there is a unique 'biggest' space–time that contains all the others starting from the same initial conditions and which is obtained by letting time go forward until the equations predict a singularity. This *maximally extended* universe is the natural candidate for the space–time that actually arises from a particular set of initial conditions, although we should remember that in principle any of the other truncated realities could be the one that exists following the initial conditions of our Universe. If the maximally extended Universe is not the extant one then the end of the Universe of space and time could indeed come at any moment 'like a thief in the night' without any observable cause or warning.

Despite all these subtleties regarding the nature of time, general relativity has failed to remove the traditional divide between laws and initial conditions. There is still always an initial slice to our space–time stack which determines what the other slices will look like.

In quantum theory the status of time is an even bigger mystery than it appeared to Newton and Einstein. If it exists in a transcendent way then it is not one of those quantities subject to the famous Uncertainty Principles of Heisenberg, but if it is defined operationally by other intrinsic aspects of a physical system then it does suffer indirectly from the restrictions imposed by quantum uncertainty. Accordingly, when one attempts to produce a quantum description of the entire Universe one might anticipate some unusual consequences for time. The most unusual has been the claim that a quantum cosmology permits us to interpret it as a description of a Universe that has been created from nothing.

The non-quantum cosmological models of general relativity may begin at a definite past moment of time defined using certain types of clock. The initial conditions that dictate the whole future behaviour of that Universe must be prescribed at that singularity. But in quantum cosmology the notion of time does not appear explicitly. Time is a construct of the matter fields and their configurations. Since we have equations which tell us something about how those configurations change as we look from one slice of space to another it would be superfluous to have a 'time' as well. This is not altogether different to the way in which a pendulum clock tells you time. The clock hands merely keep a record of how many swings the pendulum makes. There is no need to mention anything called 'time'. Likewise, in the cosmological setting we are labelling the slices in our 'space–time' stack by the matter configuration that creates the intrinsic geometry of each slice. This information about the geometry and material configuration is only available to us probabilistically in quantum theory and it is coded into something that has become known as the *wave function of the Universe*, which we shall henceforth call *W*.

The generalisation of Einstein's equations to include quantum theory is one of the great problems of modern physics. One proposed route uses an equation first found by the American physicists John A. Wheeler and Bryce De Witt. The Wheeler–De Witt equation describes the evolution of *W*. It is an adaption of Schrödinger's famous equation governing the wave function of ordinary quantum mechanics, but with the curved space attributes of general relativity incorporated as well. If we knew the present form of *W* it would tell us the probability that the observed Universe would be found to possess certain large-scale features. It is hoped that these probabilities will turn out to be strongly concentrated around particular values, in the same way that large everyday things have definite properties despite the microscopic uncertainties of quantum mechanics. If the greatly favoured values were similar to the values observed then this would give an explanation of those features as a consequence of the fact that ours was one of the most 'probable' of all possible universes. However, to do this one still requires some initial conditions for the Wheeler–De Witt equation: an initial form for the wave function of the Universe.

The most useful quantity involved in the manipulation and study of *W* is the transition function

$$T[\mathrm{x}_2, t_2; x_1, t_1]$$

This gives the probability of finding the Universe in a state labelled by x_2 at a time t_2 if it was in a state x_1 at an earlier time t_1, where the 'times' can be prescribed by some other attribute of the state of the Universe, for example, its average density (see Figure 26.3).

Of course, in non-quantum physics, the laws of Nature predict a definite future state will arise from a particular past one and we would not have need for such probabilistic notions. But in quantum physics a future state is determined only as an appropriately weighted sum over all the logically possible paths through space and time that the system could have taken. One of these paths might be the unique one but the non-quantum description would follow. We call this the *classical path*. In some cases where there exists a conventional deterministic situation, its corresponding quantum description has a transition function that is principally determined by the classical path, leaving the others to combine so as to cancel each other out, rather like the peaks and troughs of waves that are out of phase. In fact, it is a deep question whether all possible starting conditions allowed for a quantum universe can give rise to a 'classical'

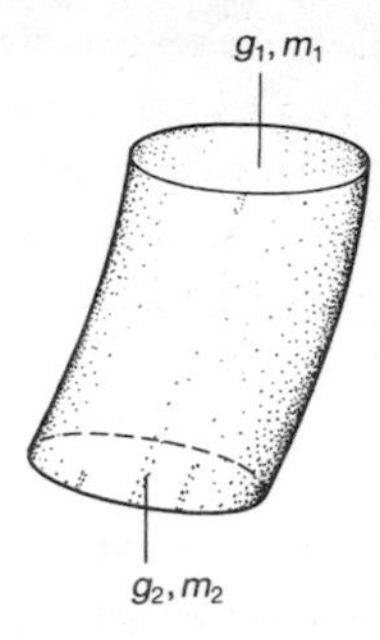

Fig. 26.3. Some paths for space–times whose boundary consists of two three-dimensional spaces of curvature g_1 and g_2, respectively, where the matter distributions are prescribed by m_1 and m_2.

universe when they expand to a large size. This may well turn out to be a very restrictive requirement, one necessary also for the existence of living observers, that marks our Universe out as unusual in the set of all possibilities. If this is true then it would also have the interesting consequence that only by a study of its cosmological consequences could a complete appreciation of quantum mechanics be arrived at.

In practice, W depends upon the configuration of the matter in the Universe on a particular slice through the space stack and upon some internal geometrical property of the slice (like its curvature) which then effectively labels its 'time' uniquely. Again, there is no special choice of geometrical quantity that is elevated above all others in labelling the slices in this way. There are many that will suffice, and the Wheeler–De Witt equation then tells you how the wave function at one value of this internally defined time is related to its form at another value of it. When we are close to the classical path these developments of the wave function in internal time are straightforward to interpret as small 'quantum corrections' to ordinary classical physics. But this is not always the case, and when the most probable path is far from the classical one then it becomes increasingly difficult to interpret the quantum evolution as occurring 'in' time in any sense. That is, the collection of space slices that the Wheeler–De Witt equation gives us does not stack naturally to look like a space–time. None the less, the transition functions can still be found. The question of the initial conditions for the wave function now becomes the quantum analogue of the search for initial conditions. The transition function slots x_1 and t_1 are where we could insert our candidates.

We have seen that the transition function T tells us about the transition from one configuration of spatial geometry, on which the matter has a particular arrangement, to another. Let us think of it as $T[m_2, g_2; m_1, t_1]$, where m labels

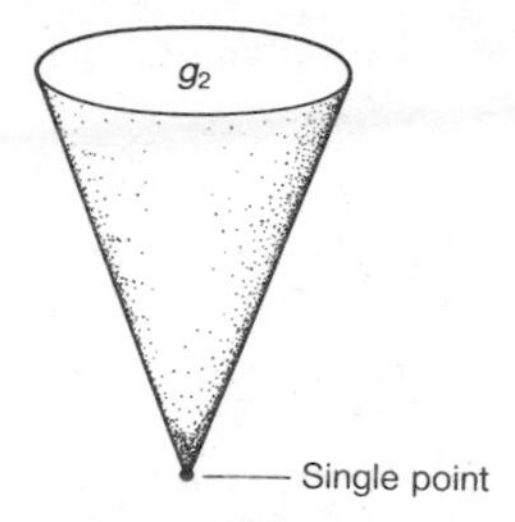

Fig. 26.4. A space–time path whose boundary consists of a curved three-dimensional space of curvature g_2 and a single initial point rather than another three-dimensional space. If there is a geometrical or physical infinity at this point then we cannot calculate the transition probability T from this point to the state with curvature g_2. If this were possible then it would give the probability of a universe with curvature g_2 arising from a 'point' rather than from 'nothing'.

the matter configuration and g is some geometrical characteristic of space, like the curvature, which we are using as an internally defined time at two values '1' and '2'. We can envisage universes that begin at a single point rather than at an initial space so that their development looks conical rather than cylindrical (as was the case in the Figure 26.3). This is illustrated schematically in Figure 26.4. Yet this is no great advance in our attempt to transform the idea of initial conditions because the singularity of the non-quantum cosmological models always shows up as a feature of the classical quantum path, and in any case we just seem to be picking a particular initial condition, which happens to describe creation from an initial pre-existent point, for no good reason. We have not severed the dualism between laws (represented here by the Wheeler–De Witt equation) and initial conditions.

There is a radical path that may now be taken. One should stress that it may well turn out to be empty of any physical significance. It is an article of faith. If we look at Figures 26.3 and 26.4, then we see how the stipulation of an initial condition g_1 relates to the state of the space further up the tube or the cone, at g_2. Could the boundaries of the configurations at g_1 *and* g_2 be combined in some way so that they describe a single smooth space that contains no nasty singularities?

We know of simple possibilities in two dimensions, like the surface of a sphere, which is smooth and free of any singular points. So we might try to conceive of the whole boundary of the four-dimensional space–time to be not g_1 and g_2 but a single smooth surface in three dimensions. This might be the surface of a sphere living in four space dimensions. One of the curious and attractive features of these smooth surfaces that mathematicians habitually consider, regardless of their dimension, which we can visualise better by returning

Fig. 26.5. An interesting path is one whose boundary is smoothly rounded off so that it consists of just a single three-dimensional space, and no sharp point at the base as there was in Fig. 4. This situation admits of an interpretation as the transition probability for creation out of 'nothing' because there is no initial state: there is only a single boundary. This can be used as the picture of the three-dimensional boundary of a four-dimensional space–time only if we suppose that time behaves as if it is another dimension of space.

to the two-dimensional surface of an ordinary sphere, is that they are finite in size but nevertheless have no edge: the surface of the sphere has a finite area (it would only require a finite amount of paint to paint it), but however one moves one never runs into an unusual point, like the apex of a cone. We might describe the sphere as being without boundary from the point of view of flat-landers living on its surface. Interestingly, such a configuration can be conceived of for the initial state of the Universe (see Figure 26.5). However—and now comes the radical step—the sphere we are using as an example is a space of three dimensions with a two-dimensional surface as a boundary. But for our quantum boundary we need a three-dimensional space as a boundary. However, this requires the four-dimensional thing of which it is the boundary to be a four-dimensional *space* and not a four-dimensional *space–time*, which is what the real Universe has always been assumed to be. Therefore, it is proposed that our ordinary concept of time is transcended in this quantum cosmological setting and becomes like another dimension of space so changing three plus one dimensions of space and time into a four-dimensional space. This is not quite as mystical as it might sound because physicists have often carried out this 'change time into space' procedure as a useful trick for solving certain problems in ordinary quantum mechanics, although they did not imagine that time was *really* like space. At the end of the calculation they just swap back into the usual interpretation of there being one dimension of time and three other qualitatively different dimensions of what we call space.

The radical character of this approach is that it regards time as being truly like space in the ultimate quantum gravitational environment of the Big Bang. As one moves far away from the beginning of the Universe, so the quantum effects start to interfere in a destructive fashion and the Universe is expected to follow the classical path with greater and greater accuracy. When this happens so the conventional notion of time as a distinct concept to that of space begins to crystallise out. Conversely, as one approaches the beginning so the conventional picture of time melts away and becomes indistinguishable from space, as the effects of the boundary conditions are felt.

This 'no boundary condition' was proposed by Jim Hartle and Stephen Hawking for aesthetic reasons. It avoids singularities from the initial state and removes the conventional dualism between laws and initial conditions. This it can achieve if the distinction between space and time is lost. More precisely, the 'no boundary' proposal stipulates that in order to work out the wave function of the Universe we compute it as the weighted aggregate of paths that are restricted to those four-dimensional spaces that possess a single, finite smooth boundary like the spherical one we have just discussed. The transition probability that this prescription provides for the production of a wave function with some other matter content m_2 in a geometrical configuration g_2 just has the form

$$T[m_2, g_2]$$

Thus there are no slots corresponding to any 'initial' state characterised by m_1 and g_1. Hence, this is often described as giving a picture of 'creation out of nothing' in which T gives the probability of a certain type of universe having been created out of nothing. The effect of the 'time becomes space' proposal is that there is no definite moment or point of creation. In more conventional quantum mechanical terms we would say that the Universe is the result of a quantum mechanical tunnelling process where it must be interpreted as having tunnelled from nothing at all. Quantum tunnelling processes, which are familiar to physicists and routinely observed, correspond to transitions that do not have a classical path.

The overall picture one gets of this type of quantum beginning is that the Wheeler–De Witt equation gives the law of Nature that describes how the wave function, W, changes. The geometry of the space can be used as a measure of time, which looks essentially like the ordinary time of general relativity when one is far from the Big Bang. But as one looks back towards that instant that we would have called the zero of time, the notion of time fades away and ultimately ceases to exist. This type of quantum universe has not always existed; it comes into being just as the classical cosmologies could, but it does not start at a Big Bang where physical quantities are infinite and where further initial conditions need to be specified. In neither case is there any information as to what it may have come into being from.

We should stress again that this is a radical proposal. It has two ingredients: the first is the 'time becomes space' proposal; the second is the addition of the 'no boundary' proposal—a single prescription for the state of the Universe that subsumes the roles of both initial equations and laws of Nature in the traditional picture. Even if one subscribes to the first ingredient there are many choices one could have used instead of the second to specify the state of a Universe that tunnels into existence out of nothing. These would all have required some additional specification of information.

The study of the wave function of the Universe is in its infancy. It will undoubtedly change in many ways before it is done. The 'no boundary' condition leaves much to be desired. It probably contains too little information to describe all the observable features of a real universe containing irregularities like galaxies. It must be supplemented by additional information about the matter fields in the Universe and how they distribute themselves. Of course, it may also be on the wrong track. The important lesson for us to draw from it here is the extent to which our traditional dualism regarding initial conditions and laws might be mistaken. It might be an artefact of our experience of a realm of Nature in which quantum effects are small. If a theory of Nature is truly unified then we might expect that it would exploit the possibility of keeping time in terms of the material contents of the Universe so as to marry together the constituents of Nature with the laws governing their change and the nature of time itself. However, we are still left with a choice as to the boundary condition that should be imposed upon some entity like the wave function of the Universe. No matter how economical its prescription it is an inescapable fact that the 'no boundary' condition and its various rivals are picked out only for aesthetic reasons. They are not demanded by the internal logical consistency of the quantum universe.

The dualistic view that initial conditions are independent of laws of Nature must be reassessed in the case of the initial conditions for the Universe as a whole. If the Universe is unique—the only logically consistent possibility—then the initial conditions are unique and become, in effect, a law of Nature themselves. This is the motivation of those who seek basic principles that might serve to delineate the initial conditions of the Universe. If this is truly the case then it introduces another new ingredient into our thinking about the Universe, because it points to a fundamental asymmetry between the past and the future in the make-up of the laws of Nature. On the other hand, if we believe that there are many possible universes—indeed may actually *be* many possible universes 'somewhere'—then initial conditions need have no special status. They could be just as in more mundane physical problems: those defining characteristics that specify one particular actuality from a general class of possibilities.

The traditional view that initial conditions are for the theologians and evolution equations for the physicists seems to have been overthrown—at least temporarily. Cosmologists now engage in the study of initial conditions to discover whether there exists a 'law' of initial conditions, of which the 'no boundary' proposal would be just one possible example. This is radical indeed, but perhaps it is not radical enough. It is worrying that so many of the concepts and ideas being used in the modern mathematical description—'creation out of nothing', 'time coming into being with the Universe'—are just refined images of rather traditional human intuitions and categories of thought. Surely, it is these traditional notions that motivate many of the concepts that are searched for and even found within modern theories that are cased in mathematical form. The 'time becomes space' proposal is the one truly radical element that we cannot detect as the inheritance of past generations of human thinking in philosophical theology. One suspects that a good many more habitual concepts may need to be transformed before the true picture begins to emerge.

CHAPTER 27

Universe began in no time at all

The belief in immortality rests not very much on the hope of going on. Few of us want to do that, but we would like very much to begin again.

Heywood Broun

The hands on my clock tell me what time it is, but not what time is. Time rules our lives from cradle to grave without ever fully revealing itself. We sense its passing: slowly whilst sitting in the dentist's chair; unexpectedly quickly when having a good time or working at a computer keyboard.

In Chester Cathedral there is an epitaph that captures this impression we share about the flow of time:

> For when I was a babe and wept and slept, Time crept;
> When I was a boy and laughed and talked, Time walked;
> Then when the years saw me a man, Time ran;
> But as I older grew, Time flew.

Such subjective feelings about time's flow, beautiful though they may be, are clearly unreliable. They differ from person to person and even vary with the number of alcoholic beverages you have imbibed.

Scientists have always tried to do better than psychologists and poets at telling what time is. They suspect that it is more than simply God's way of stopping everything from happening at once. But they are most at home with the more mundane business of finding ever more accurate ways to measure time. The most accurate clocks count oscillations of atoms. Their rates do not depend upon the vagaries of the outside world. For while you can measure time with an eggtimer, your measurement will not be the same as anyone else's. Their eggtimer may have a different shape of glass bowl filled with more finely ground sand that slides faster down the glass. Even pendulum clocks keep different time from place to place on the Earth's surface. The pull of gravity that keeps the pendulum swinging decreases slightly as one approaches the Equator where the effects of the Earth's rotation are greatest. To overcome this variability our most accurate time standards use 'ticks' of a clock which correspond to natural vibrations of atoms. Astronomical observations of distant stars reveal that atoms in space vibrate in the same way as on Earth. They are governed by universal laws and constants of Nature. We could use atomic clocks to tell (in some suitable language) a railway commuter on Alpha Centauri what we mean when we say that his train is going to be 'one hour' late. 'One hour' is just our term for a certain very large number of atomic vibrations.

About 150 years ago, humanity woke up to the importance of time. Local change, whether daily, seasonal or annual, had always been appreciated, but for thousands of years the wider role of time had been ignored. During the second half of the nineteenth century, the ubiquity of time at last began to emerge. Darwin's theory of evolution revealed the importance of time in the development of living things. The discovery of the laws of thermodynamics, which describe the degradation of energy into disordered forms, led astronomers, philosophers and theologians to contemplate the eschatological bleakness of a Universe steadily unwinding into a state of increasing disorder and uninhabitability: a future universe unimaginably different from the past. Geologists recognised the tides of change that had ebbed and flowed deep in the Earth's crust over aeons of time, and astronomers began to study the formation of our solar system from an earlier state of spinning gas and dust. Then, to cap it all, in the 1920s astronomers discovered that the entire Universe is expanding. Distant galaxies are rushing away from each other and from us: the whole cosmic environment is constantly changing. Look into the past and the galaxies will be closer together; look back further still and there will be no galaxies at all, just a host of dismembered atoms bathed in a sea of microwaves that we see today with NASA's COBE satellite. Keep backtracking and we come to an apparent beginning of the Universe about 15 billion years ago. The hint of a beginning to the Universe—the Big Bang—at a finite time in the past has led cosmologists to reconsider the whole question of the meaning, as opposed to the mere measurement, of time. For if the Universe is everything there is—all space, all matter and all time—then time comes into being along with everything else at the Big Bang. The Universe is created with time, not in time.

This is an ancient viewpoint that St Augustine espoused long ago for entirely theological reasons. Today, we see it endorsed by Einstein's theory of general relativity. As a consequence, lots of common-sense questions about the beginning of the Universe cease to have meaning in any deep sense. 'What happened before the beginning of the Universe?' sounds a sensible question but it derives from a notion that the Universe is a 'thing' that suddenly appeared *in* time. But there is no transcendental time sitting 'outside' the Universe. It is easy to think that because every event or thing that we see had a cause that the Universe must also have necessarily had a cause. But the Universe is neither a 'thing' nor an 'event' in the ordinary sense. It is the totality of all events and things (and no doubt more besides) and may not need a cause. For all members of a society have a mother, but that does not mean that society itself had a mother. A natural implication of this view of time in the Universe is to ask whether there really is some deep aspect of reality called 'time' at all. If time is bound up with the nature and contents of the Universe perhaps it is just another name for something else.

This game of the name is something we play all the time. When I look at my grandfather clock and talk about the passage of 'time', I am talking about counting swings of the pendulum in space. I do not see anything called 'time' directly. Look at your digital watch and what you call 'time' is an accounting of vibrations of an atom. As one goes back to the beginning of the Universe the temperature and density of material within it gets so large that no structures survive that could be used as clocks. One keeps time using some aspects of the Universe's overall make-up, like the density of matter within it.

In recent years, cosmologists have speculated that as the apparent beginning of the Universe is approached we may find, not the beginning of time, but a state in which time ceases to exist. The widely publicised conjecture of Stephen Hawking and his American collaborator, Jim Hartle, is that time becomes just another form of space in the extreme environment of the Big Bang. By contrast, as the Universe expands and cools the impression of 'time' would become more vivid and persistent. As we trace history back to its first fleeting instants, we may find that the notion of time melts away like a will o' the wisp leaving us with a story of the Universe that begins with the words 'Once upon a time there was no time'. Only time will tell whether these ideas survive the detailed scrutiny they are now being given by physicists.

If they are right then the fact that we live late in the Universe's history, when the impression of time is clear, is no coincidence. Complex living things like ourselves must be made from elements like carbon, nitrogen and oxygen. These elements are produced by a process of nuclear alchemy inside the stars. Over billions of years hydrogen and helium gases are burnt into carbon, nitrogen and oxygen. The stars explode and distribute the elements of life through space. Eventually they end up in you and me. So any Universe containing the building blocks required for the evolution of biological complexity must be billions of years old. Remarkably, since the Universe is expanding, this also means that Universes that contain life must be billions of light years in size. You do, indeed, need 'world enough and time' to think about the Universe after all.

CHAPTER 28

Arrows and some banana skins

Humpty Dumpty sat on a wall
Humpty Dumpty had a great fall;
All the King's horses and all the King's men
Couldn't put Humpty back together again.

Nursery Rhyme

'Time flies like an arrow; fruit flies like a banana', according to the Marx brothers' analysis of the vexed question of the arrow of time. This temporal conspiracy at the heart of things has many subplots: why do we have memories of the past, why is the Universe expanding rather than contracting, why do measures of disorder grow with time in closed systems, while the equations of physics seem to possess an entirely democratic attitude with respect to the direction of time?

This book is an attempt to grapple with the different problems of the arrow of time with a high degree of analytical care. Separate chapters deal with time-asymmetric systems in elementary particle physics, the reduction of the wave function and the problem of time in quantum mechanics, advanced and retarded radiation, but there is a special focus of interest upon cosmology as the crux of the arrow of time problem. For the most part the author's treatment is wide-ranging and substantial. The discussion is carefully signposted and chapters possess point-by-point summaries of the argument and the author's principal conclusions. The author announces at the start that he adopts the 'block universe' view of space–time as an objective entity, as opposed to imagining that time is just an artefact of a special human perspective on cosmic events. He is keen to uncover errors of reasoning that have infested many recent analyses of the origins of the arrow of time in physical problems, for example attempts to distinguish initial and final states of closed universes using

---

A review of *Time's arrow and Archimedes' point: new directions for the physics of time* by Huw Price, Oxford University Press, 1996

quantum cosmological boundary conditions, attempts to do the same using inflation, and fallacies in the use of the Wheeler–Feynman absorber theory. In these aims he succeeds with great clarity. His arguments are accessible to physicist and philosopher alike. However, moving to the more general cosmological claims, there should be surprise at the way in which some cosmological speculations are adopted as fact and used as the basis for the discussion of the cosmological time–arrow problem.

The author makes great play of his analysis of cosmological arrows of time, in particular the oft-asked question of why the entropy of the Universe increases. He claims that this problem is misconceived in some sense: that the real problem is not to explain the trend but 'to explain the low-entropy past' of the Universe. The discussion of cosmology revolves around this issue. The problem with his approach is twofold: first, we do not have any reliable measure of 'the entropy of the Universe'; (nor do we even know that such a quantity exists); secondly, we do not know that the Universe began in a low-entropy state as the author assumes. In fact, one might claim that the situation is even more fundamentally uncertain: there is no way that we can *ever* know whether or not the Universe began in a low-entropy state.

The problem of specifying the entropy of the Universe is that we do not know all the potential contributors to it. There are classical and quantum contributions, dynamical contributions, particle species, and so on. However, there is no reason to think that the entropy of even the visible part of the Universe is fully measured by counting the massive and massless particles it contains or by simply calculating something like the Bekenstein–Hawking entropy of a black hole the size of the particle horizon. Neither of these measures keep track of all the forms of disorder that can distinguish one universe from another. They do not have the properties we would expect of measures of cosmological disorder and information. In order to obtain a measure of the Universe's entropy we need to solve the great problems posed by the study of generic forms of complexity: what is complexity and how do we quantify in a way that tells us whether one thing is more complex than another. At present there are many candidates, each satisfactory for some types of complexity, but none appropriate for all.

More problematic than the question of a measure of the entropy of the Universe is the assumption that the entropy of the Universe was low in the distant past (at 'the beginning' perhaps?). It is not clear where the author has taken this key datum from. I know of no sure evidence for it even if we pick on simple intuitive measures of cosmological non-uniformity to measure its entropy. Nor could there be any such evidence. Our observations of the Universe only allow us to sample its structure on and close to the inside of our past light cone. Observations of the microwave background, radiation isotropy, and deductions from primordial nucleosynthesis allow us to go back some way in time but still leave us very far from determining the character of the beginning of the expansion. Even with perfect observations we could learn only about the structure of a minuscule part of the structure of the Universe at a moment of cosmic time close to the apparent beginning. All the information we would need to determine whether the whole initial state was of low entropy is not causally accessible to us even with perfect observations (neutrino telescopes, graviton telescopes) because of the finite speed of light. Price only seems to be able to overcome this limitation by assuming what is to be proved.

The inflationary universe theory, which is consistent with current observations and should be decisively tested by satellite-born instruments over the next decade, opens up a Pandora's box of further possibilities and probabilities, which makes it even harder to defend the assumption that the Universe began in a state of very low entropy. Inflationary universes can begin in highly irregular states of high entropy, far from equilibrium, and end up appearing very regular to observers like ourselves. Different regions can undergo different amounts of inflation (which may even continue for all future time like a stochastic branching process). Our visible universe (a finite part of the possibly infinite, highly in homogeneous, whole) is all that we can see. The entropy of the whole Universe is not accessible to us. Indeed, we might expect to be a fluctuation with special features that permit expansion to last long enough to create the conditions in which the evolution of complexity is possible. Inflation predicts just that, but we can never test that prediction fully.

For these reasons, despite the fact that the author has done physicists a great service in laying out so clearly and critically the nature of the various time–asymmetry problems of physics, I have reservations about the author's discussion of the cosmological arrow of time. They are really only discussions of particular speculative assumptions about cosmological initial conditions. In grappling with the symmetries and asymmetries of cosmological time we still seem to be nearer the end of the beginning than the beginning of the end.

Ingram Pinn

Part 8

Quantum reality

I hate reality, but it is still the only place where I can get a decent steak.

Woody Allen

A well-known theoretical physicist once told me of an embarrassing experience he had during his first week as a graduate student at an American university. Like most new graduate students in theoretical physics, he had been an extremely good undergraduate student, scoring highly on anything that looked remotely like mathematical physics, and with a special talent for solving well-posed problems in double quick time (usually about six of them in three hours). Bursting with confidence, he decided to attend the first gathering of members of the department that occurred each week just before the seminar, where everyone chatted and drank tea. Our student found himself sitting next to a short, balding scientist of European extraction who seemed to be muttering something quietly to himself, but as the student sat down his thoughtful neighbour turned to him and said dreamily, 'I don't understand quantum mechanics'. Seizing his chance to help, the student replied 'Oh, it's quite straightforward. It goes like this . . .' and spent the next five minutes giving an elementary tutorial to his listener about how you write down the basic equations, find simple solutions, and all the other techniques you learn as a student. The older scientist listened with exquisite politeness, now and then complimenting him on the clarity of his explanation, but also sometimes asking an awkward question that he couldn't quite answer. Soon the departmental chairman clapped his hands and announced that everyone needed to move on to the lecture room. At this point the older scientist thanked the young student profusely and said, 'I am so pleased to have met you. What is your name? I am Eugene Wigner'. The student began to hope that the floor would open up beneath him as he realised that his tutee had been one of the greatest theoretical physicists of the 20th century, Nobel prize-winner, and, worst of all, one of the greatest contributors to the development of quantum theory. Our student had unfortunately discovered that he was also one of the most modest scientists that one could ever meet.

So, what was Wigner talking to himself about? It was concerning one of the great mysteries of science. Quantum mechanics provides us with a mathematical equation that allows us to predict the workings of the atomic and molecular realms with fantastic accuracy. Some predictions of this theory have been found to be correct to more than 12 decimal places. Yet the physical interpretation of the mathematical procedures that lead to this unprecedented accuracy have been the subject of argument ever since the theory was formulated. 'Nobody', as Richard Feynman once asserted, 'understands quantum mechanics!'

As cosmologists have faced up to the fact that they need to apply quantum theory to the entire Universe if they are to probe the earliest moments of the Universe's history, so the mystery has deepened. An array of popular books has attempted to explain the problems that quantum theory creates for the philosopher, whilst physicists divide into different camps according to which interpretation of quantum mechanics they favour. Until quite recently, it was believed that these different interpretations were nothing more than different ways of saying the same thing and no experiment could ever distinguish them. Recently, it has been suggested that this might well not be the case and the differences might not be merely semantic. The next chapter, '*In the best of all possible worlds*', is a review of Paul Davies' excellent account of the problems of interpreting quantum mechanics. It is followed by another review, this time of Abraham Pais' encyclopaedic review of Niels Bohr's life and work.

Bohr was one of the founders of quantum mechanics and led an eventful life. He was smuggled out of occupied Denmark during World War II in order to escape the Nazis and eventually ended up in America working on the atomic bomb project. Ask physicists who they would choose as the second greatest scientist of the 20th century and they would overwhelmingly vote for Bohr. As a personality, and in working style, he could not have been more different than Einstein. In his deepest thinking about the nature of measurement he framed the concept of complementarity which he believed to be important in many other areas of human thinking. It captured the unavoidable contribution of the act of measurement to what is recorded by any process of measurement. Michael Riordan has described the quantum picture of subatomic reality by analogy as being 'a lot like that of a rainbow, whose position is defined only relative to an observer. This is not an objective property of the rainbow-in-itself but involves such subjective elements as the observer's own position. Like the rainbow, a subatomic particle becomes fully "real" only through the process of measurement.'

The next review, '*Complementary concepts*', is of a book by John Honner that traces the origins of Bohr's concept of complementary, showing how it had its roots in a favourite work of speculative fiction and in his father's physiological studies. The last, '*Much Ado About Nothing*' is about the problem of 'nothing'—to which there is a good deal more than meets the eye. Quantum theory teaches us that the vacuum is not just nothing. It is a sea of ghostly particles, continually appearing and disappearing. Under very delicate circumstances their presence can be revealed by their minute effects upon the energy levels of atoms. The quantum vacuum is not defined by the absence of anything, but by virtue of it being a state where energy is minimised. Later, we shall see that it is this potential for the vacuum to possess changeable properties, some of which are more likely than others, that may provide an explanation for many of the properties of the astronomical universe around us.

CHAPTER 29

In the best of all possible worlds

Historians have concluded that Heisenberg must have been contemplating his love life when he discovered the Uncertainty Principle:—When he had the time, he didn't have the energy and,—when the moment was right, he couldn't figure out the position.

Tryggvi Emilsson

It has been said that the mark of good philosophy is to begin with an observation so mundane that it is regarded as trivial and from it deduce a conclusion so extraordinary that no one can believe it. In this latest addition to his string of successful explanatory texts, Paul Davies has chosen an area of physics that bears this hallmark more clearly than any other. Quantum physics, bristles with extraordinary conclusions. It has grown out of the need to explain a series of unusual experimental results concerning the behaviour of atomic and radiative processes. The traditional approach towards building a model of the atom had assumed that Nature was economical in its use of design principles, choosing merely to scale down the scheme of the solar system to atomic size rather than pick a completely different structure. Unfortunately, this simple view leads to a flagrant contradiction with our everyday observation of the world. If the atom is imagined to be nothing more than a mini solar system, held together by the balance between electric forces and the motion of electrons encircling its central nucleus, then a serious problem emerges.

Electrons possess a quality familiar to us in the large, electric charge. For the charged electrons to be maintained in orbit around the nucleus they must do work and so expend energy. Their speed will therefore drop and so they will move a little closer to the nucleus. However, just like the pirouetting ice-skater, the electron will now spin even faster, lose energy still faster, approach the nucleus again, and so on. This slippery road, what physicists call an 'instability',

---

A review of *Other worlds: space, superspace and the quantum universe,* by Paul Davies, Dent 1980

means that the electrons would be doomed to spiral into the nucleus in a very short time. Atoms could not exist.

In 1913, Niels Bohr suggested a simple, but profound modification to the naive 'solar system' type of model which completely resolves this difficulty. He proposed that the energy of electrons bound within an atom is restricted to certain discrete values; multiples of a fixed and fundamental energy 'quantum'. The electrons cannot change their energies continuously and so do not steadily slide into the nucleus. In order to lower their energy they have to change its value by a whole quantum. This new model of the microworld, so different from our intuitive idea that one ought to be able to put an electron in orbit anywhere about a nucleus, was a remarkable success. It explained a host of observational anomalies in spectroscopy and provided a sound basis for the systematic study of all atomic and molecular phenomena. Some years later Erwin Schrödinger discovered a differential equation which enables the quantization of energy to be understood more clearly. Furthermore, it revealed the reason why sub-atomic phenomena simultaneously display the properties of waves and discrete particles.

All the information available about an object can be embodied in a mathematical construction, termed its 'wave function'. If we know this function then Schrödinger's equation allows us to predict the behaviour of the object as far as it is possible for it to be predicted. It also allows us to calculate the only quantity of immediate relevance to us as observers: the *probability* that the object we measure will possess any particular configuration when we examine it. In some cases the probability will be so overwhelmingly high for a particular realization, that it amounts to a certainty. In that case the familiar 'classical' (non-quantum) picture of the world is an excellent approximation to the quantum model.

The key difference between the classical and quantum pictures is that, although both are built upon deterministic differential equations, in the latter the quantity whose evolution is exactly determined (the wave function) is not observable. When we carry out a measurement of a system we inevitably alter the form of this wave function. All the wave function and the laws of physics can give us are probabilitistic connections between subsequent measurements of the system. This apparent uncertainty is not interpreted as a deficiency of the quantum theory as currently formulated, but as an intrinsic feature of the perturbation which the very act of measurement introduces.

Dirac, writing soon after these features were first realized, succinctly summarized the impact of this discovery: 'The classical tradition had been to consider the world to be an association of observable objects moving according to definite laws of force, so that one could form a mental picture of the whole scheme. It has become increasingly evident, however, that Nature works on a different plan. Her fundamental laws do not govern the world as it appears in our picture in any direct way, but instead they control a substratum of which we cannot form a mental picture without introducing irrelevancies.'

Other worlds opens with a simple, non-mathematical introduction to the ideas and consequences of quantum physics. The author's aim is to emphasize the new world view this physical theory forces us to contemplate. He lays a good deal of emphasis upon the unpredictable and apparently chaotic picture of microscopic processes. It might have been illuminating to show that there is another side of the coin: Bohr's quantum theory only allows atomic electrons to occupy certain discrete orbits around the nucleus. (In modern terms we would say that an electron would most probably be observed in these orbital states.) Now suppose Nature were built upon non-quantum principles: one could place a single electron about a proton in an infinite number of stable orbits just by choosing their relative velocity and separation to produce a balance between Coulomb and centripetal forces. The result—every hydrogen atom would be different: no replicating systems would be possible and the price of microscopic exactitude would be chaos in the large. Seen in this light, quantization also ensures that atomic systems do not continuously change under the constant buffeting of neighbouring atoms and photons. It is the basis for the large-scale stability of nature.

The early chapters of *Other worlds* revolve around three famous paradoxes spawned by the existence of the wave function. These display the alarming subtleties of interpreting quantum theory. Although familiar to practising physicists, the dilemma of 'Schrödinger's cat', 'Wigner's friend' and the 'Einstein–Podolsky–Rosen paradox' do not seem to have tunnelled their way into the public imagination to the same extent as relativistic conundrums like the 'twin paradox'. Yet, taken at face value, they are equally bizarre. The culprit in these challenges to common sense is inevitably the 'wave function'. It is always possible to set up a quantum mechanical state which is a combination of any two other states. However, one is at liberty to choose these two component states so that they correspond to situations which, on a *macroscopic* view, are mutually exclusive. In the case of the 'cat paradox' the two states are chosen to represent 'dead cat' and 'live cat', respectively. So how do we interpret the quantum state which looks like the average of a live and dead cat?

Paul Davies skilfully employs these paradoxes to paint a very clear picture of the strange notion of reality physicists have been forced to accept. Yet, in spite of his lucid presentation, there is a problem he never addresses, and which will be lurking in the minds of his audience: Is not quantum theory, in spite of its overwhelming operational success, after all just a model, a picture of reality? It is strongly supported by observation; but then so was Newton's theory of gravitation. To build a philosophy or interpretation of reality upon the particular representation our minds have chosen as the vehicle of physical description is a dangerous project. And then again, the reader wonders, if our rationality is purely a manifestation of such microphysical uncertainty, what grounds can we possibly have for faith in its deductions?

The latter half of the book moves on from a discussion of specific problems of quantum measurement to outline the so-called 'many worlds' interpretation of quantum mechanics. Parmenides was the first to introduce the idea that the subjectivity of every man's experience introduces a non-uniqueness into our ultimate descriptions of the world. The 'many worlds' interpretation of quantum theory, introduced by Hugh Everett in 1957, develops this idea in a modern mathematical form. Since its inception it has been a constant area of dispute among eminent physicists. The 'many worlds' interpretation stands in complete contrast to Bohr's traditional interpretation of quantum mechanics. It takes the mathematical formalism at face value and restores the wave function to a status that admits it as a true description of the world. When pushed to its logical conclusion Everett's interpretation maintains that all possible outcomes of measurements actually exist. Each elementary process, each observation and pointer measurement we make, picks a path through the mesh of all possible outcomes. This splitting process that measurement induces in the world of possibilities is unobservable, and all the unrealized 'other worlds' are disconnected from each other and from our own experience. At first sight this appears like a guide to solipsism extracted from a dusty tome of scholastic philosophy, but, in context, it possesses many appealing features which its supporters claim ameliorate problems in the interpretation of quantum theory. The 'other worlds' lead, in a natural way, to the statistical properties of quantum systems. Microscopic indeterminacy is seen as a consequence of our access to such a limited portion of the 'superspace' of all possible worlds. The evolution in the 'superspace' is entirely deterministic.

Davies describes how the 'many worlds' picture can be used to motivate consideration of an entire ensemble of possible universes (multiverses?), the conditions and behaviour of which fill out the whole infinite range of possibilities. They can all be interpreted as necessary and extant mutations from the world we experience rather than purely hypothetical constructions. Now it becomes to reformulate the ancient notion of 'the best of all possible worlds' in a framework that demands the dissimilar alternatives actually coexist! It should not surprise us to learn that our own Universe belongs to a subset of the 'superspace' of universes which possess life-supporting features like stability, large size, low temperature, carbon chemistry, and so forth. But, turned upon its head, this tautological statement creates an interesting perspective on Nature: if we observe the Universe to possess structural features which are, *a priori*, remarkable but necessary for the evolution of intelligent observers, we should temper our surprise at these observations. This view of many superficially peculiar features of our own universe's large-scale structure has gained in popularity among cosmologists in the past ten years and goes by the name of the 'anthropic principle'. It is well described in chapter eight of Davies' book.

If one subscribes to the 'many worlds' interpretation then all possible worlds really do exist, sitting like the volumes of Borges' *Library of Babel*—all possible typographical permutations of all possible books. Like the librarians assiduously searching for the intelligible works, we can examine the spectrum of structural possibilities in the ensemble of worlds and evaluate the special features of our own in the exclusive society of life-supporting universes. Davies discusses how one could imagine the size of 'superspace' enlarged to encompass candidate worlds with different values for the fundamental constants (for example, changing the ratio of the proton to electron mass). Surprisingly, these perturbations from our own universe invariably seem to be still-born, unable to generate the basic atomic structures necessary to evolve complex, replicating organisms. They seem doomed to be devoid of observers. Fine tuning between the conditions required to support life and those actually observed is interpreted as 'natural selection' in 'superspace'.

If one does not take the 'many worlds' picture seriously (and after all, there may exist infinitely many worlds in which even Everett himself does not believe it) then subscribers to the anthropic principle may regard our own universe as unique and pre-programmed to produce 'observers'. This approach is closely related to a train of thinking with a long and chequered history. From the dawn of history philosophers have argued for the existence of some form of teleological structure working within, or transcending, the natural world. This view has traditionally been supported by 'design arguments' which point to the beneficent and meticulous anthropocentric plan of Nature—a design that superficially operates with the existence and continued survival of humans in mind. It is a view that many great scientists of the past have subscribed to. Following Cicero and Aquinas it was generally construed to support an argument for the existence of a benevolent deity, or deities.

Not surprisingly, there have been many enthusiastic opponents of this 'natural theology' which has evolved and adapted itself to existing scientific knowledge for many hundreds of years. Yet, it was a world-view that was scarcely perturbed by the logical critique of Hume and Kant or the satirical tirade of writers like Montaigne and Voltaire. 'Philosophical' objections had little impact because they lacked the observational basis of the simple design arguments of Derham, Ray, or Paley which were firmly welded to the observation of Nature and acted as an important stimulus to its investigation. The teleological antecedents of the 'anthropic principle' were erected upon all manner of remarkable natural contrivances, each apparently tuned to the pre-conditions of our survival. Living organisms clearly fitted their environments hand-in-glove; and, to some, even laws of Nature—the inverse-square law of gravity and the dynamics of the solar system—appeared to be miraculously preferred amid the sea of anthropocentrically unfavourable possibilities. Although Darwin was to revise our perspective on many of these contrivances in a radical fashion, he confessed that he had been strongly influenced by the classical design arguments which so clearly placed before him the striking examples of adaptation that are exhibited in Nature.

The 'anthropic principle' that Paul Davies describes can be seen as a sophisticated version of the early design arguments. Instead of appealing to contrivance and fitness of local and transient features of our own universe, it points to those invariant or unique aspects of its global structure which are well adapted to accommodate the evolution of living organisms. The incorporation of the Everett 'many worlds' interpretation of quantum mechanics generates the other 'universes' necessary to validate any statements of comparative reference. Again the reader may object that this is just a convenient way of seeing things. After all, we recall another attempt to place the notion of 'the best of all possible worlds' on a rigorous footing. Maupertius interpreted his principle of 'least action' in this vein, claiming it to be a superlative argument for the existence of God. The possible motions of a dynamical system can be parametrized by a quantity called the 'action'. The path actually taken by the system is always that which minimizes the action; and this provides an elegant way of deriving its equations of motion. Viewed only in this light the action principles of physics appear entirely teleological, and Maupertius interpreted their existence as evidence that the actual course of Nature was preferred amid a sea of mathematically precise but unfavourable alternatives. The least-action path was the best of all possible worlds. However, action principles can easily be reformulated in terms of deterministic differential equations which evolve specific initial conditions forward in space and time; the teleological aspect vanishes. We have alternative representations of a physical problem which, although mathematically equivalent, are metaphysically divergent.

These are some of the issues that Paul Davies's stimulating and provocative book touches upon in his presentation of some standard and speculative aspects of modern quantum theory. His exposition should appeal to a wide audience of interested laymen and they will find no more lucidly written introduction to an area that has challenged some of the twentieth century's greatest physicists for an interpretation.

CHAPTER 30

Great Dane

A philosopher once said "It is necessary for the very existence of science that the same conditions always produce the same results". Well, they do not.

Richard Feynman

'There were giants on the earth in those days'. And indeed there were. Two men tower head and shoulders above all other 20th-century physicists through the breadth and depth of their achievements. One was lucid, quotable, persuasive, and peripatetic; the other, complex, obscure, misunderstood, living and working almost entirely in the land of his birth. One was Einstein. The other was Bohr. Whilst almost every soul on the face of the Earth has heard of Albert Einstein, few outside the halls of science have heard of Niels Bohr. Yet whilst we owe our understanding of the Universe in the large to the insight of Einstein, it was Niels Bohr who first untangled the complexities of microscopic matter and knitted them into a coherent pattern that revealed the true depths of meaning that lie trammelled up within the inner space of the atom and its nucleus. Within these small worlds a curious legislation holds sway that forbids us learning of its state with ever-improving accuracy. No matter how perfect our instruments of observation, there exists an irreducible uncertainty in our simultaneous determination of certain 'complementary' pairs of its properties. The act of intervening to observe a state produces an inevitable and indeterminable change in its structure that suffices to destroy the Cartesian picture of an 'observer' of the world separated from the 'observed' like a birdwatcher in a perfect hide. We cannot determine the location and the speed of motion of a particle at the same time with complete precision. This indefinite 'quantum' reality that holds sway in the microworld of the elementary constituents of matter is now the bread and butter of all university courses of physics; it dictates the workings of the manifold of electronic wonders that ease the burdens of everyday life. The correspondence between its predictions and our observations of the world are of staggering accuracy. It is the most accurate and

A review of *Neils Bohr's Times*, by Abraham Pais, Oxford University Press, 1993

most successful description of any aspect of the natural world that we possess. As the sizes of objects increase from the subatomic to that of the everyday things around us, so the effects of quantum ambiguity become less and less obvious and the statistical predictions of the quantum description of the world converge upon the familiar 'classical' picture first discovered by Newton and his followers 300 years ago.

No phenomenon in the Universe has ever been observed that fails to obey the predictions of the quantum mechanics that Bohr established to order the workings of the atomic world. But this success story was hard won. The unravelling of the mystery of the quantum interpretation of matter involved one of the greatest changes in thinking that scientists have ever been forced to come to terms with. The story of how that new perspective emerged and transmogrified the physicists' world view is at once the story of physics in the first half of the 20th century and the story of Niels Bohr.

Abraham Pais is a physicist who has made fundamental contributions to the quantum picture of reality. But during the last 14 years he has established himself as a historian of science who combines deep scientific insight with personal knowledge and meticulous scholarship. In modern times he is perhaps the unique example of a world-class physicist turned historian. His first excursion into the history of science was his acclaimed life of Einstein. Whilst that volume was widely acclaimed by scientists, it was not so enthusiastically received by some historians who, whilst not denying Pais's mastery of the facts, looked in vain for a deeper interpretation of them which made contact with the character of Einstein himself and the dramatically changing world that his career spanned. In this volume and in similar style Pais brings his insight to bear upon a subject even closer to his heart than Einstein. For Bohr was Pais's mentor for many years when he worked in Copenhagen as one of Bohr's young research assistants, living for long periods in the Bohr's family home, and joining in the intellectual struggle to unravel the secrets of the atom and the nucleus within it. There is, of course, a danger here. Some may feel that Pais is too close to Bohr and his family to provide a full and impartial account of his life.

Niels Bohr was born in Copenhagen in 1885. He grew up in a large and happy family and was throughout his life particularly close to his brother Harald who was both a world-class mathematician and a footballer of international calibre. Niels himself was a keen footballer and played in goal for a Danish club side. Their father was a physiologist of no mean achievement, having been nominated for the Nobel prize in medicine in both 1907 and 1908 for his discoveries concerning the effect of carbon dioxide upon human blood. Niels' parents and his teachers soon recognised his extraordinary abilities and we find him displaying extraordinary physical intuition and powers of analysis right from his student days. The influence of his father was perhaps an important but subtle one. For Christian Bohr had interesting philosophical views regarding 'complementarity' in science and human enquiry. He had seen that there were indeed questions which one could ask of the world which, once asked, render certain other questions meaningless. The biologist must kill an animal in order to discover its internal structure and so the nature of life could not be investigated by means of certain questions. The dichotomy of free will and determinism seemed to him to be another pair of complementary opposites. Later, Niels would use something akin to this notion of complementarity to deal with the interpretation of our observations of the quantum world. In the latter part of his life, when his interests broadened to include politics and philosophy, he would return to the philosophical implications of the existence of complementary attributes of the world and the extent to which an analysis of any given situation needed to take into account the entire state of affairs—the observer and the observed—in order to avoid contradictions and paradox. One can even find his early espousal of an evolutionary epistemology in which our view of the world is seen to be a consequence of a process of natural selection that selects for those images that are in accord with the true nature of the things in themselves. Thus 'realism' with respect to our scientific picture of those aspects of physical reality whose accurate apprehension is a necessary prerequisite for our own evolution is justified.

Pais' account of Bohr is enriched by the author's first-hand experience of living and working in Bohr's home for long periods of time. There emerges a picture that is confirmed by others with that same experience—the successful marriage and the pleasure that he took in all his six children, one of whom was to follow in his father's footsteps to become a Nobel prize-winner in physics, and the shattering double blow he was dealt by the death before his own eyes of his oldest son in a boating accident and the death of the youngest from meningitis, four years later, when only ten years old.

The author's subtitle reveals the threefold division of his life that he sees in retrospect. His work as a philosopher is small and certainly had no effect upon leading philosophers during and after his lifetime. On this part of his work Pais is least helpful and has not engaged in a detailed analysis of its content and significance. Instead, he has devoted himself to the other two arms of Bohr's triangle of activity. For his physics is the story of the growth of modern physics: the end of classical physics and the beginnings of the quantum theory and its successes in unravelling the structure and vibrations of atoms, the mathematical understanding of the periodic table of the elements, and the creation of modern chemistry with the means to predict and dictate the properties of atoms and molecules. Hand in hand with these successful applications of the new quantum mechanics Bohr was wrestling with the question of its correct interpretation and the fact that it gives only probabilistic predictions of the behaviour of the world. In this quest he entered into a long-standing debate with Einstein regarding the correctness of the quantum description of reality. Einstein regarded quantum mechanics as an unsatisfactory but effective description of how things happen but whose lack of determinism would not be part of the ultimate theory of the world. Einstein pursued this belief by the creation of 'thought-experiments' which he believed exhibited logical contradictions created by quantum mechanics. In every case Einstein was wrong and Bohr impressively unravelled the true state of affairs in all of Einstein's examples, convincing him of the self-consistency of the quantum description of events when correctly and fully applied. In these interactions between Bohr and Einstein, and in his discussions with other great physicists, one is struck by his remarkable intuition and speed of thought in unfamiliar territory. Some found it altogether too much; James Franck even left Bohr's circle to take up a professorship in America because, despite his great love and respect for Bohr, he found working close to him unsettling: 'Bohr did not allow me to think through whatever I did to the end. I made some experiments. And when I told Bohr about it, then he said immediately what might be wrong, what might be right. And it was so quick that after a time I felt that I am unable to think at all . . . Bohr's genius was so superior.'

With the establishment of a self-consistent interpretation of the mathematical workings of quantum mechanics, Bohr's next great quest was the understanding of the nucleus of the atom, a problem that he was motivated to attack by George Gamow's discovery that 'alpha' radioactivity could be understood as a quantum mechanical 'tunnelling' process of a sort that possesses no classical analogue. Gamow visited Bohr's Copenhagen Institute during the years 1928 to 1931 and used some of that time to write the first book on theoretical nuclear physics. Pais tells the story of how Bohr's theoretical research in nuclear physics led to his inevitable involvement in the wartime quest for its military exploitation. Here, many readers may be a little disappointed because Pais completed his work before the declassification and release of the records of the German atomic bomb project and the debriefing of German scientists, including Heisenberg and von Weiszäcker, in England following the two nuclear explosions over Japan. But neither those accounts nor Pais's are able to shed very much light on the famous private meeting that took place between Bohr and Heisenberg in Copenhagen whilst the city was under German control. It is not clear whether Heisenberg was seeking to enrol Bohr in the German bomb project or merely seeking to know of Bohr's current interests and intentions. Heisenberg's own relationship with the German authorities is also ambiguous. Soon afterwards Bohr escaped to Sweden, from whence he was flown to Britain under somewhat bizarre circumstances. He made the flight lodged in the bomb-bay of a Mosquito. He was instructed in the use of the oxygen mask in preparation for the high-altitude phase of the flight prior to take-off. Presumably, his thoughts were elsewhere because when the time came to don the breathing equipment the pilot failed to raise him on the intercom; reducing altitude, he landed in Scotland to discover Bohr in good shape claiming that he had slept soundly all the way—evidently unconscious due to a lack of oxygen.

Bohr's arrival in Britain and his subsequent voyage to the United States to contribute to the Manhattan Project marked the start of the third phase of his life: his political campaigning for international cooperation. Bohr was one of the first potentially influential advocates of glasnost in the arena of weapons-related scientific expertise. He believed that the sharing of nuclear know-how would create a more stable environment than the fast-developing scenario of a secret arms race between rival superpowers. As far as I am aware, there have been no detailed evaluations of the effectiveness (if any) of Bohr's interventions in this respect by political historians of the period. All the commentators seem, like Pais, to be physicists whose opinion of Bohr is so coloured by their high regard for his other intellectual achievements that one wonders about the objectivity of their analyses. Certainly, we know that Churchill had little time for Bohr's arguments and their meeting was less than a success for Bohr: rumours remain that Churchill's reaction to the obscure and unnecessarily labrythine arguments of Bohr being 'get this blithering idiot out of here' and subsequently Bohr was regarded as a security risk because of his advocacy of information sharing with the Soviet Union. As a direct result of his interview with Churchill his attempts to influence Roosevelt were doomed to be ignored and following the meeting of the two world leaders in New York their *aide de memoire* contains an item that 'enquiries should be made regarding the activities of Professor Bohr and steps should be taken to ensure that he is responsible for no leakage of information, particularly to the Russians.'

The reasons for Bohr's hostile reception, the imperviousness of Churchill to his reasoning are interesting issues that Pais does not dwell on. Perhaps there was no more evidence to be gathered and no other witnesses of those post-war events to question? But I suspect not. From Pais's account of those years one merely gets a general feeling that Bohr was a victim of the very same philosophy that he had so powerfully stressed in the analysis of the events in the quantum world—that one must consider the system as a whole and not divide it into a dichotomy of the observer and the observed. The ideas that he was promulgating were not illogical or absurd in themselves but they have to be considered contextually. In the immediate post-war years of Stalinist rule in Eastern Europe, following the most appalling of all world wars, the time was not ripe for his idealist suggestions. They were unstable with respect to those who would seek to take information whilst holding some back. The means simply did not exist for verification of weapons agreements. Today, Bohr's arguments seem timely and almost uncontroversial. One might be tempted to say that he was ahead of his time. But in politics, being ahead of the times can be as disastrous as it can be prescient.

Niels Bohr was a thinker of enormous depth and subtlety. Abraham Pais has laid before the reader a masterly ordering of the facts. From them he has drawn a portrait that is clear and authoritative in those areas close to Pais's own heart: Bohr's brilliant physics and his deep humanity nurtured by the closeness of his family and his paternal affection for the young physicists who passed through his school seeking his intuition and wisdom to further the search for the secrets that lie at the heart of matter. Yet the wider range of Bohr's interests, his excursions into philosophy, and the reverberations of his attempts to bind the nations into peaceful coexistence will repay further study by others without the influences that personal acquaintance and scientific appreciation inevitably bring to bear. In these days of the wide popularisation of science and scientists there is much loose journalism about 'new Einsteins' but if Pais's two great biographies of Einstein and Bohr teach us anything it is that in breadth and depth of achievement and insight, no living physicist approaches these two stars in the firmament of 20th century physics.

CHAPTER 31

Complementary concepts

The opposite of a correct statement is a false statement. But the opposite of a profound truth may well be another profound truth.

Niels Bohr

Harald Bohr was once asked why his writing and lecturing could display such clarity while that of his brother Niels was a miasma of obscurity. He replied that 'everything I say follows exactly from what I have just said, but everything which Niels says follows exactly from what he is about to say'. Thus it is entirely fitting that Niels Bohr's theory of quantum reality should produce such a challenge to our traditional belief in determinism. It swept away the Cartesian dualists' picture of the scientist as a bird-watcher holed-up in a perfect hide, able to report upon the world as it really is so long as he exercises a little care in handling his field-glasses.

Bohr's legacy is that there can be no such bird-watcher and no such hide. Everything in nature—whether it be temporarily playing the role of the observer or of the observed—is irreducibly linked. The act of measurement plays an essential, inseparable and unpredictable role in determining the result of that measurement. It is not simply that the act of measurement disturbs the object of measurement in some way, so that if only we were a little less ham-fisted we could determine both where an elementary particle was and how it is moving; Bohr claimed that the success of quantum mechanics as a description of the world proved that no such undisturbed reality exists. The uncertainty principles of Heisenberg thus tell us the minimum degree to which we can decouple the observer from the observed, or the limit beyond which naive realism cannot be pushed. They provide us with pairs of so-called 'complementary' quantities which cannot be measured simultaneously with arbitrary precision because they are merely coarse, classical notions which cannot coexist in the microcosm of the quantum world.

Such ideas and the challenge of interpreting the process of quantum measure-

---

A review of *The description of Nature: Niels Bohr and the philosophy of quantum physics,* by John Honner, Clarendon Press, 1988

ment provide both physicists and philosophers with technical problems that still await a convincing resolution. But John Honner has avoided a further discussion of these problems and chosen instead to explain Bohr's philosophy of Nature in general and his doctrine of complementarity in particular. Bohr regarded this doctrine as having far wider applicability than in the quantum measurement problem, but towards the end of his life remarked that 'I think it would be reasonable to say that no man who is called a philosopher really understands what is meant by the complementary descriptions.' Whether this was the fault of the philosophers or the result of Bohr's singular inability to express himself clearly and finally, the reader will be in a better position to judge after reading Honner's guide to Bohr's philosophical thought, *The description of nature*. Honner abstracts what he believes to be Bohr's definition of complementarity in the following terms: 'Our position as observers in a domain of experience where unambiguous application of concepts depends essentially on conditions of observation demands the use of complementary descriptions if the description is to be exhaustive.'

Bohr picked upon the elements of traditional philosophical conflicts like that of determinism versus free will as candidates for complementary concepts in the sense that the adoption of one of these viewpoints renders the other meaningless. But he never managed to interest philosophers in this quest and one might have hoped that Honner would give a clear reason for this failure. Unfortunately, he does not. While he pays careful attention to both published and unpublished sources of Bohr's work, this account lacks clear signposts and conclusions. It will be a valuable guide to the growing number of scholars interested in unravelling Bohr's thought, but it falls just short of clarifying complex arguments. The book's strong point is the dissection of individual arguments and claims so they can be read and evaluated in skeletal form, but it is less successful in weaving them together into a single tapestry. This may not be entirely the author's failing. One can ask whether the elusive nature of Bohr's ideas is not an intrinsic part of their message—I am sure that he would have maintained that clarity and accuracy are complementary and hence, in some circumstances, mutually exclusive concepts.

Regrettably, one essential element in the story which does not even receive a mention is Bohr's family background. For this appears to be the origin of his predilection for seeking complementarity here, there and everywhere. His father was a professor of physiology at the University of Copenhagen and involved in the nineteenth-century quest of the teleomechanists to harmonize vitalistic and mechanistic modes of explanation in the natural world. He first introduced the notion of complementary explanation to assuage these problems and engaged in discussions with philosophical friends like Hoffding on a range of related epistemological problems. Indeed, later in life Niels recalled his father's picture of complementary explanation in the realm of physiology and continued correspondence on the subject of complementarity with Hoffding. On this important area Honner is silent, yet one would like to know in detail just what Bohr did gain from his father and what other influences helped fashion his unusual approach to the world.

After reading this interesting survey of a very complex thinker one is left wondering as to how it could be that Bohr and Einstein, the two creative giants of twentieth-century physics, could have been so different. Where Einstein was almost childlike in his simplicity of visualization, eminently quotable, clear, engaged always in the reduction of the complex to the simple, with a deep faith in the ultimate correctness of a mathematical description of the world and a definite observer-independent reality, Bohr was difficult and obscure, attracted by paradox, in search of the ineffable rather than the simple, at once distrustful of mathematics as a vehicle of ultimate explanation and unwilling to admit the existence of any unobserved reality. Yet to Einstein we owe our understanding of the world of outer space where gravity holds sway and to Bohr our understanding of the sub-atomic microcosm of inner space.

CHAPTER 32

Much ado about nothing

Among the great things which are found among us the existence of Nothing is the greatest.

Leonardo da Vinci

'Nothing', according to the *Encyclopaedia of philosophy*, 'is an awe-inspiring yet essentially undigested concept, highly esteemed by writers of a mystical or existentialist tendency, but by most others regarded with anxiety, nausea, or panic.' At root, it owes its perennial philosophical popularity to an inevitable juxtaposition of opposites: 'why is there something rather than nothing?' By this false dichotomy and our incessant desire to explain why there is indeed 'something' we have been wrong-footed into regarding 'nothing' as somehow the more natural state of affairs.

Ancient philosophers were wont to make very strong statements about Nature's abhorrence for nothingness when it can have 'being' instead; but not so ancient philosophers honed devices to deal with the problem of the vacuum's actual existence. In retrospect they all seem rather unsatisfactory. For the Platonist the only true vacuum has a 'pi' in the sky existence in another world of pure ideas. We see only an imperfect representation of this timeless entity which is presumably, therefore, necessarily a non-vacuum. For the operationalist the meaning of the vacuum is the sequence of steps one must take to construct one; but this is an unattainable, for in the real world the act of intervention destroys the very concept one seeks to construct. Finally, for the empiricist, all that remains is a description of the most rarified state that one can possibly create and this leaves us with a variety of vacuum defined by James Clerk Maxwell in the nineteenth century as 'the vacuum is that which is left in a vessel after we have removed everything which we can remove from it.'

Whereas philosophers have worried greatly about the presence and role of mind in the definition and explanation of the properties of the world, scientists, for the most part, have been happy to treat the presence of sentient

---

A review of *The philosophy of the vacuum*, edited by S. Saunders and H.R. Brown, Clarendon Press, Oxford, 1991

beings or their mechanical agents as harmless irrelevancies when it comes to getting to the heart of things. Thus Newtonian physics makes no mention of observers and we can happily talk about a Universe that contains nothing but space and time: neither matter nor motion. The first chink in this pristine image was perhaps Maxwell's conception of a microscopic 'intelligent demon', who, by identifying faster than average molecules, could cause a gas to heat up in defiance of the second law of thermodynamics. Suddenly, the presence of an observer could challenge the laws of physics. Later, the invention of the quantum theory by Bohr, Heisenberg and Dirac threatened to do away with the vacuum as a physical concept altogether. Heisenberg's Uncertainty Principle revealed the intimate connection between the observer and the observed and suggested to some that there could be no phenomenon unless it was an observed phenomenon. The very act of measurement produces an irreducible effect upon the system being measured so that complete knowledge of one attribute of the system expunges all knowledge of some other 'complementary' attribute. The consequences of this for students of nothingness were far-reaching. Before quantum aspects of reality were recognised one could conceive of the vacuum as being simply 'nothing' pure and simple, but not afterwards. The limitation upon our specification of the state of a physical system imposed by Heisenberg's Principle prevents us from making a sure statement that there is nothing inside our box. Clearly, the notion of a state in which no particles or radiation are present is not a very useful one when it comes to describing the microscopic quantum world. As an alternative to the traditional one, physicists moved towards the idea of the vacuum as the lowest energy state of the system. This has all sorts of interesting consequences: the vacuum state can change with time, differ from place to place and have a variety of attributes; moreover, uniqueness need not be one of them.

The simplest picture of the new vacuum that emerged through the work of Schwinger, Feynman, Dyson and Tomonaga could not be in greater contrast to the traditional image of 'nothing'. It was a sea of particles and antiparticles, continually appearing and disappearing but each pair being too ephemeral to allow observation to be made of them without violating Heisenberg's Principle. At first one might think that we can do without such excess metaphysical baggage in physics, but, remarkably, this sea of 'virtual' particle–antiparticle pairs was predicted to produce tiny changes in the observed energy levels of atoms. These were subsequently observed in high-precision experiments.

In recent years the study of how the vacuum state of the matter in the Universe behaves during its earliest moments has become a focus of activity amongst cosmologists. Many theories of how matter behaves at very high temperatures, like those near the Big Bang, predict that matter may find itself temporarily caught in a 'false' vacuum state. This is an equilibrium, like a pencil tottering on its point, that is unstable, and will change into another stable 'true' vacuum state, lying flat on the table. The transit between the two states involves the transfer of 'potential' energy into energy of motion. In the early Universe the transit between the two states results in the creation of a peculiar tension in the state of matter that will temporarily accelerate the expansion of the Universe at a rate far in excess of that expected in the standard picture of the early history of the Universe. This period of accelerated expansion, or 'inflation', as it has become known, is remarkable because it predicts that the visible universe should possess a number of the peculiar properties that cosmologists have long sought to explain.

There is thus a good deal more to the vacuum than one might have innocently expected. Although the content of the vacuum provides important material for philosophers of science to ponder, they are ill-equipped to take the necessary ideas on board. Most have some familiarity with the conundrums presented by special relativity and quantum mechanics. But that is pre-war physics in many respects. The modern physics of the vacuum requires at least some familiarity with quantum field theory and the general theory of relativity and the mathematical complexities of both those subjects are formidable. Outsiders need an introduction.

This book grew out of a short conference held in Oxford in 1987. Some of the articles were first presented as expository papers at that gathering, others have been written later in order to complement them, while the opening article by Einstein is a little known essay plucked from the archive of his papers at the Hebrew University. The aim has been to draw together philosophically minded physicists to expound the modern concept of the vacuum and the problems it still presents. The expositors are distinguished and wide-ranging in their talents and coverage. There are gravitation specialists like Penrose and Sciama, mathematicians like Atiyah and Braam and theoretical physicists, like Aitchison, Hiley and Finkelstein, with a wide cross-section of views. They are intermingled with philosophers like Saunders, Weingard and Brown, who are hard to distinguish from physicists. The resulting long-lived fluctuation from the vacuum that the authors' contributions have created is a valuable one but perhaps not in the way that the editors imagined. The level of mathematics and physics required to appreciate more than the opening pages of most of the contributions will place this work off limits to most philosophers of science. On the other hand, physicists will find it extremely interesting. It covers technical subjects in an accessible way intermediate to that found in popular articles and reviews for the cognescenti. Moreover, it spans several periods of interest in different pictures of high-energy physics and gives a linked overview spanning the hiatuses between them. The main criticism from experts will be that the time-lag between conception and delivery of this volume has also spanned a period of dramatic change in the way physicists view the most elementary structure of matter and space–time. 'String' theories threaten to displace quantum field theories from their central position amongst the venerated paradigms of physics and any well-rounded overview of the vacuum needs to introduce these new ideas a little more systematically and accessibly than has happened here. This particular lack is part of a wider omission of more introductory survey articles in the volume. A few of the authors have not given sufficient thought to the likely readership and have launched into over-technical expositions that would be more at home in the pages of the *Physical Review*. But, that said, the philosopher of science should take on board the implications of this book and its degree of difficulty. There is no reason why the most interesting problems on the interface between physics and philosophy should be any simpler than those at the interface between philosophy and mathematical logic. Those wishing to pursue them need to acclimatize themselves to a far higher level of mathematical physics than has been their habit in the past. If one wants to write about 'being and nothingness' then one must discover first that a revolution has occurred in our pictures of both which can so easily pass you by because the quantities involved happen to have kept the same names in the process. For those with the necessary expertise this book will provide an illuminating and authorative exposition of a subject that is so many-sided that its dimensions are positively fractal. Nothing will never be the same again.

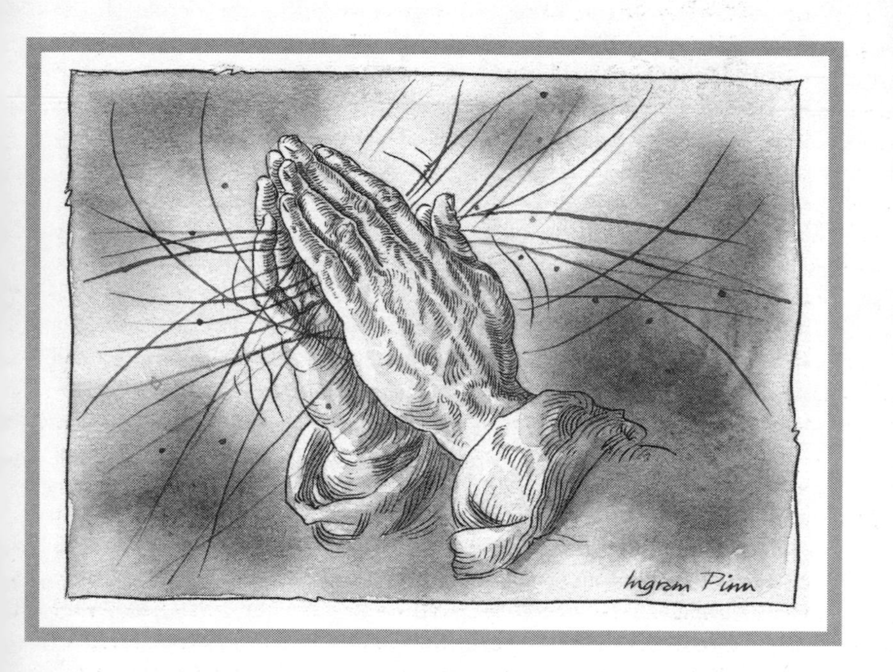
Ingram Pinn

Part 9

Religion and science

Without a parable modern physics speaks not to the multitudes

C.S. Lewis

The relationship between religion and science is far from obvious. For some, science is a device to 'prove' religion, while for others it is a device to prove the very opposite. But the term 'science' is almost as big a catch-all as 'religion'. In reality, different sciences display as much heterogeneity as different religions. The sciences of the very large, astronomy, cosmology, and fundamental physics, appear more sympathetic to religious traditions; while the life sciences, biology, psychology, and anthropology, appear antagonistic. There is a little more to this divide than simply the distance of the object of scientific study from the workings of the human mind and the factors that fashion its beliefs. It reminds us of the divide that we highlighted in earlier chapters, between the laws of Nature and their outcomes. Those sciences that have much contact with the fundamental laws of Nature lay greater emphasis upon the symmetries and simplicities of Nature than those that concentrate upon the outcomes of those laws, where the symmetries are broken or disguised. As a result, those who work with the laws of Nature are more likely to be impressed by the idea of a 'mind' behind the appearances, a God who is the master mathematician, than are the students of symmetry breaking. For the latter, the world is a complicated outcome of a process of sifting and selection, often influenced by chaotic unpredictability. These two ways of looking at Nature can also be found at the root of attempts to argue that the wonderful harmonies of Nature are evidence for a grand designer. There is a very old collection of arguments that point to the way in which living things are seemingly tailor-made for their environments and how the parts of their bodies are so well constructed to carry out their functions: a favourite Victorian example was the fine optical engineering of the human eye. This is an argument about the fortuitous interrelationships between the outcomes to the laws of Nature. It was swept away by Darwin's discovery that a process of natural selection which favours outcomes that promote survival, will in the long run ensure that those outcomes become more widespread in a population. By contrast, there was a less ancient form of design argument, first made popular by Newton and his supporters 300 years ago, which pointed to the remarkable simplicity of the forms of the laws of Nature. Were they to be slightly changed then life would become an impossibility. Modern forms of the anthropic principle show that this argument allows many, more complex, conclusions to be drawn. However, logically, it is perfectly consistent with design; it suffers from none of the defects of the argument from the fortuitous arrangements of the outcomes of Nature's laws, and is entirely unscathed by Darwin's discoveries.

There is quite another lesson to be drawn, by analogy, from this dichotomy between symmetrical laws and asymmetrical outcomes. It teaches us that appearances can be deceptive. Any attempt to divine the ultimate nature of the Universe or its immaterial purposes from appearances is going to be vastly (perhaps infinitely) more difficult than understanding how its physical structure is related to its underlying laws. For, should we not expect that the appearances will hide or break the 'symmetries' that underlie the nature of the immaterial and transcendental?

This is a problem that theologians and philosophers have perhaps not fully appreciated. They are well aware of the gap that may exist between the true nature of things and the appearances they present to our minds. Immanuel Kant taught us that 200 years ago. But physics teaches us that there may exist another gap: that the outworkings of a principle need not themselves maintain that principle. Sometimes you have to be cruel to be kind.

The chapters in this section discuss some of these issues, '*Doubting castles in the air*', looks at the apparent differences in attitude that religious and scientific thinking seems to require: the first demands faith while the second encourages doubt. It goes on to look at some of the areas of fundamental physics, and the Platonic interpretation of mathematics favoured by many physicists, in the light of religious attitudes. The second chapter is a short review of a book by John Polkinghorne, one of the clearest commentators on the relationship between religion and science. Polkinghorne is particularly keen to exploit discoveries in the study of unpredictability and chaos to arrive at a fuller understanding of traditional theological questions about providence. '*Accepting Pascal's wager*' is a review of a most unusual book by Stephen Brams. The author is a well-known expert on the theory of games. In this context, a 'game' is any situation where two or more players are in competition with each other. The players can be people playing poker, a childrens' game like paper–scissors–stone, voters choosing candidates, or animals competing for natural resources in different ways. In each case, the players have different strategies open to them. Each results in a certain payoff, positive or negative, according to the strategy adopted by the other players. No player knows which strategy the other will adopt for the next play. Suppose the game is played over and over again. A number of questions arise: is there a best possible strategy that you should adopt in the sense that any deviation from it will see you worse off in the long run? In the game paper–scissors–stone, for example, the best strategy is to play each with equal frequency (one in three) in the long run regardless of what your opponent plays; you cannot expect a better payoff in the long run from any other strategy.

Brams is one of the world's experts on this approach to the analysis of conflicts in politics and international affairs. Here, he applies his expertise in a novel way. He asks whether there is any evidence from our knowledge of things that we are in effect playing a game with a supreme being who is adopting an optimal strategy. In particular, are things like the problems of free will or suffering understandable as optimal strategies for a particular type of game? Are there games where the optimal strategy is logically undecideable?

This type of game-theoretic theology began hundreds of years ago with the great French mathematician and philosopher, Blaise Pascal, who argued that atheism was illogical because the optimal strategy to play against God when considering His existence was to believe in it. The potential infinite positive payoff to be received from being right is so great compared with the relatively minor cost of holding such a belief if God does not exist or the infinitely negative payoff if you are wrong and He does exist. This is called Pascal's wager. Pascal, by the way, was a devout theist. This was not an argument that his own faith rested upon.

The Christian tradition that the Universe was created out of nothing has been much discussed in recent years after new quantum cosmological theories provided ways to talk about the probability that a Universe appears out of nothing. In '*Getting something from nothing*', I describe some of the background to this question, together with some unusual properties of our Universe that seem to remove some of the barriers to its appearance out of the physicists' nothing, to which there is a good deal more than meets the eye.

CHAPTER 33

Doubting castles in the air

Religions die when they are proved true. Science is the record of dead religions.

Oscar Wilde

Scientists are much like other folk. Yet, in modern folklore they have come to be seen as the gatekeepers of truth: holders of the touchstone against which all other systems of knowledge and belief must be tested and tried. Those found wanting are dispatched to the vacuum of meaninglessness. Most other disciplines have adapted to survive by adopting at least a veneer of scientific paraphernalia so as to appear respectable. But the most striking contrast to the belief system of science is that of 'religion' in its many traditional forms. Once it held the same authority that science commands today. Yet even today it has important things to say about many of the subjects within the dominion of scientific investigation. As a result there exists a steady trickle of books and articles about 'science and religion'. Leaving aside the crankier variety, most are quasi-apologetic, aiming to show that religion and science are not mutually exclusive. They provide historical examples of Christian scientists or highlight the larger role played by monotheistic religious beliefs in providing a world-view in which the notion of science could germinate and grow. Whereas a century ago they would have been matched in number by strident opposition to any theistic view of the world, these opposing voices have grown strangely silent, not, one suspects, because they have admitted defeat, but because the gradual secularisation of modern thought has made the religious alternative appear an insignificant threat to those who strongly oppose it.

The interaction between scientific and religious beliefs about the world is a deeply complex one, involving a weighing of many historical and psychological factors. There are many contrasts between the approaches to the world that these belief systems offer but if one were to be forced to pick only one to highlight then the issue of doubt versus certainty is the most interesting and provocative.

Scientists are trained doubters and such a trepidacious perspective has been enshrined within some philosophies of the scientific method that emphasise the experimental testing of hypotheses and the ephemera of scientific paradigms. Some, like Popper, wish to accept as meaningful only those claims that submit themselves to the possibility of falsification. The pros and cons of such a rigidly human-centred criterion of meaning have been debated fairly thoroughly by scientists and philosophers of all persuasions. An awkward dilemma is the fact that there exist highly respectable 'scientific' theories, like 'atomism'—the view that matter is composed of a hierarchy of substructures culminating in some basic building blocks that we tend to call elementary particles—which have an ambiguous pedigree. This view of the nature of matter has been tested in certain ways by a multitude of experiments and is taught in all the universities of the world. But its origins are entirely unscientific. It arose amongst certain schools of ancient Greek thought as an entirely metaphysical or religious idea for which there was not a shred of observational or experimental evidence; indeed, it would be thousands of years before the means would even exist to search for such evidence. Thus, ancient atomism would have failed any contemporary criterion for scientific validity, yet it contained a kernel of deep truth about the world. We might wonder how many similar ideas about the world are judged devoid of meaning because they happen to fail some popular contemporary criterion of human creation.

Let us return to the question of doubt or certainty. Whereas scientists do well to be sceptical, 'believers' of any religious persuasion maintain a core of beliefs which they regard as unfalsifiable or transcending falsification. In effect they must hold that there exists a realm of 'absolute truth' which humans have a means of gaining limited entry to by way of a very strait gate. The means by which such access is gained involves elements that are not amenable to complete rationalization. For the beginning student of science this sounds a lame and fantastic claim. But it is instructive to compare it with the claims that many notable scientists (for example, and most recently, Roger Penrose in *The emperor's new mind*) make concerning the relationship between mathematics and the world. In order to reconcile the extraordinary fact that mathematics works as a description of the physical world they choose to pursue a Platonic view that mathematicians do not merely invent mathematics to suit their own purposes: they discover it. This requires them to maintain that there exists another world of mathematical entities or ideas and the mathematical nature of reality is a manifestation of the blueprints of absolute truth that reside there. It also requires them to suppose that there exists some strange means by which we are able to interact with this other world of mathematical ideas so that our minds become aware of them. The means by which we do so is left unexplained. A surprisingly large number of mathematical scientists subscribe to the Platonic view that the world *is* mathematical in this sense. The greatest of all 20th-century logicians, Kurt Gödel, believed that the mind had some means of accessing the nether world of mathematical reality that did not involve physical sensation but resulted in the human acquisition of mathematical intuition. But, despite these remarkable consequences, few scientists recognise 'pi in the sky' mathematical Platonism as religious view. Indeed, less seriously, one could go further, out of deference to the perenially popular undecidability theorem of Gödel, and claim that if a religion is defined to be a system that contains unprovable statements then not only is mathematics a religion but it is the only religion able to prove itself to be one.

There have been times when religion and science were closely allied and there have been times when they appear to have been in open conflict. The contrast between doubt and certainty sheds some light upon this history and also upon the current resurgence of common interests at the interface between theology and fundamental science.

The cogs of the Newtonian world view which meshed 300 years ago provided onlookers with a picture of the world as a vast mechanism following God-given 'laws that never shall be broken. For their guidance hath He made.' Whereas earlier attempts to create a natural theology by gleaning evidence for the worldly activities of a benign creator had focused upon particular fortuitous outworkings of Nature, like the human eye or any tailor-made animal habitat, the Newtonian apologists pointed to the existence of the invariant laws themselves as evidence for an omnipotent lawgiver behind the scenes. We must appreciate, however, that this scientific picture of the world was not treated like any modern one. It was not doubted in any way. Newton was widely regarded as having discovered how the Almighty had constructed the world: 'Nature and Nature's laws lay hid in night: God said, "Let Newton be" and all was light.' Hence the close study of, and sympathy with, such scientific pictures of the world did not entice the adherent to adopt a sceptical point of view which might then have dangerous consequences if let loose elsewhere. Scepticism was engendered by metaphysics not by physics.

One of the reasons that the Newtonian picture of the world was taken to represent a piece of absolute truth about the world was because it was founded upon the ancient principle of geometry. Newton's *Principia* is a tour de force of the power of Euclidean geometry. In and before Newton's time, and for more than a century and a half afterwards, theologians were able to point to the existence of Euclid's geometrical theory as a true description of the world—the human discovery of a little piece of ultimate and absolute truth. Its existence enabled one to refute any sceptical claim that ultimate truths like those sought by religious believers were beyond human ken.

In the 19th century this state of affairs changed in a dramatic way. A number of continental mathematicians discovered that there exist other logically consistent non-Euclidean geometries which arise if the famous 'fifth' parallel postulate of Euclid is not assumed. This gave rise to geometries on curved rather than flat surfaces. Triangulation was a well-defined operation on these surfaces but gave rise to 'triangles' whose interior angles did not sum to one hundred and eighty degrees. All this sounds rather mundane unless one is a geometer, but the wider consequences for religious and philosophical thought were dramatic. Something of a crisis arose which traditionalists responded to by seeking reasons why the old geometry of Euclid merited a special status. These efforts succeeded only in identifying a distinction without a difference. Worse still, Einstein would eventually replace Newton's rectilinear world of inflexible space and immutable time by one that revealed real space and time to follow the curvilinear meanderings of the non-Euclidean geometries. Faced with an infinite sea of logically valid alternatives, no longer could one point to Euclidean geometry as a unique part of the ultimate truth about reality. Its status was downgraded to that of but one amongst many possible geometries, some of which now looked like man-made systems.

Many sceptics and iconoclastic thinkers seized upon these discoveries to challenge the assumption of absolute truth in a host of different areas of human thought and practice. No longer did they 'hold these truths to be self-evident' whether they concern the rights of man or right-angled triangles. Whereas once, the assumption that there existed 'best' systems of ethics and government had seemed a reasonable one to entertain by analogy with the status of Euclidean geometry, now, the relative status of such notions was an obvious parallel to the new position of Euclidean geometry. Strikingly, the term 'non-Euclideanism' came to signify any relativist challenge to a doctrine of God-given truth across the entire spectrum of human activity. Articles appeared promulgating 'non-Euclidean economics' and 'non-Euclidean systems of government'. Later, the discovery of new logics would carry this trend further to undermine the assumption that classical logic was an absolute truth which simulated the action of perfect human thought. Finally, Tarski would go on to banish any well-defined notion of absolute truth from the game of logic.

All this relativism in science and mathematics arrived at a time when the effect of Darwin's assault on another citadel of absolutism was still being absorbed. Natural selection revealed that many of our physical attributes, like those of the other less cerebral members of the living world, owed their nature to a gradual process of change rather than a once and for all design plan. These events led to an awkward separation of religion and science which was easy to portray as a state of outright warfare. The events of this cold war are often told, retold, and reinterpreted. They assume a misleading stature, when viewed from afar by the casual observer, of the interaction between science and religion.

Subsequently, we find a curious long-term trend in the relationship between the theological emphasis upon certainty in the face of the scientific habit of provisionalism that we now find enshrined in semi-popular philosophies of science like that of Kuhn, with its never-ending cycle of revolution and revision. Where the discoveries of non-Euclidean geometries and new logics had placed emphasis upon the unexceptional nature of the truths that man had fished intuitively from the great ocean of possibilities, the new discoveries of 20th century physics took a new turn. Quantum theory introduced the notion of an absolute limit to our understanding of the world and hence opened up a new front for the religious scientist to probe. The God-of-the-gaps might find new lacunae to tunnel through under the cloak of quantum uncertainty. Such an apologetic persists in some quarters even today and has become allied to a search for inevitable gaps in human accounts of the physical world occasioned by the possibility of chaotic behaviour. In our inability to determine the future states of chaotic systems some apologists find room for the controlling influence of a God-of-the-gaps. Yet clearly this indeterminism is not intrinsic. It is merely our inability to determine in practice unless we look down at the quantum level whereupon we recover nothing new at all—merely intrinsic quantum uncertainty.

Recent theological interest in the development of fundamental physics in its quest to uncover a 'Theory of Everything' is understandable. Many of the questions that the two subjects consider—the creation of the Universe, the nature of time, the end of the Universe—are identical. But most interesting is the way in which this quest once again offers the seductive possibility of absolute truth to those who had become accustomed only to mathematical 'models', which provide only the latest human edition of the truth. In recent years certain types of physical theory founded upon mathematical symmetry and the requirement of self-consistency have been unexpectedly successful in narrowing down the possible theories that could simultaneously describe all the forces of Nature as different manifestations of a single unified force. Some optimists talk of the work of fundamental physics being complete if only one of these theories should prove acceptable and then be confirmed by experiment. Many particle physicists regard such theories not simply as models or approximations to reality but as exact descriptions of a reality that is for some unknown reason intrinsically and Platonically mathematical at its deepest level. For the theologian something akin to absolute truth re-emerges to replace the lost exemplars of classical logic and Euclidean geometry. And indeed we see that if the pious hope of the scientists for a single all-embracing theory of the laws of Nature is successful, then that success will have been to turn fundamental science into an unfalsifiable collection of statements about the world founded upon a faith in the primacy of symmetry and mathematics.

CHAPTER 34

Immortal, invisible . . . not really there?

The only thing that bothered me was that there were too many plus signs.

Paul Erdös, after a visit to a Catholic college.

Christian apologetics, like those of other religions, bend with the winds of change. As the dominant paradigms that we use to encapsulate the Universe have evolved so has the emphasis of the apologists' arguments. From the beginnings of serious philosophical thought you can extract two strands of thinking about the world: one emphasises the unchanging elements of and behind the physical world as pre-eminent; the other points to the ephemera of individual events as the overriding reality.

We see this divide in ancient philosophy where the Platonists tried to associate the true nature of things with the unchanging forms behind the world of observation, while the earthiness of Heraclitus and Aristotle heralded the individual outcomes as the vital element. There have been parallel strands in theology that have set the unchangeableness of God—'the same yesterday, today and forever'—against the process theologians' picture of a continuously changing and evolving deity 'perfecting all things unto himself'.

In physical science there is a similar dichotomy between laws and events. Laws of change can be represented as invariances of particular 'conserved' quantities in Nature, which are equivalent to the preservation of some underlying symmetries in the way of the world. The outcomes of those laws, on the other hand, need not respect that symmetry and will in general be far more complicated in form than the laws themselves. Physicists talk endlessly about the symmetry and simplicity of the laws of Nature but invariably fail to impress the life scientists whose bread and butter is the complicated outcomes of the laws of Nature that result from the higgledy-piggledy of natural selection where the underlying symmetries are totally hidden.

Science has been dominated by the strand that emphasises the unchanging

A review of *Science and providence,* by John Polkinghorne, *New Science Library*, 1991

laws and symmetries of Nature as her most indelible hallmark from the time of Newton until little more than a decade ago. Religious apologetics followed this lead, pointing to the harmonious form and unerring logic of the laws of Nature —the 'laws that never shall be broken. For their guidance hath He made'— rather than the peculiar appropriateness of their asymmetrical outcomes—the eye, the hand, the match between creature and habitat that natural selection so elegantly explained—as the signature of the author of nature.

But the emergence of the sciences of complexity has changed the course of the river of change. There exist complex phenomena whose evolution through time cannot be represented by any abbreviated formula, so that the most succinct representation of their information content is nothing less than their complete history. There exists no briefer encapsulation: no invariant whose preservation is equivalent to their ever-novel behaviour.

The focus upon these complex and chaotic processes has urged us to regard 'becoming' as irreducible to mere 'being': we cannot expunge the concept of time from all of the everyday phenomena that we encounter by restating them as abbreviated statements of invariance. Professor John Polkinghorne takes this change of focus within science very seriously and attempts to extract from it some new theological lessons. His short book is based upon a series of lectures delivered on Oxford in 1987, which was probably a little too early to take on board the most striking developments in the study of complexity. Although the author writes eloquently upon topics from the problem of evil to miracles, from prayer to time and providence, he has been greatly impressed by these developments. He repeatedly uses them in his attempts to create a new theological perspective.

Here, Polkinghorne claims that, because chaotic systems possess so strong a dependence upon their initial states that they are unpredictable, there is scope for the deity to make his presence felt in an otherwise deterministic world. This idea spreads its influence through the discussion of providence and miracles.

He argues: 'The generation of weather is a much more complex process, within which it is conceivable that small triggers could generate large effects. Thus prayer for rain does not seem totally ruled out of court. In this way one can gain some rough comprehension of the range of imminent action. It will always lie hidden in those complexes whose precarious balance makes them insusceptible to prediction. The recently gained understanding of the distinction between physical systems which exhibit being and those which exhibit becoming may be seen as a pale reflection of the theological dialectic of God's transcendence and God's imminence . . . If God acts in the world through influencing the evolution of complex systems, he does not do so by the creative input of energy.' The author sees this breach of traditional determinism as the new gap through which God can slip.

While I believe that this change of emphasis in the scientific view of the world should have important consequences for theological and philosophical ideas about the nature of things, I think one might be sceptical as to whether Polkinghorne has picked a fruitful point to exploit. If one restricts attention to the Newtonian, deterministic view of the world, and ignores quantum mechanics for a moment, then while it is true that there exist complex and chaotic phenomena whose behaviour depends with exponential sensitivity upon their starting state, this means simply that they are indeterministic in practice (for us), not that they are indeterministic in principle. There is no logical gap for God to sneak through and direct events without doing violence to the observed laws of Nature unless the deterministic laws are temporarily suspended. If we go along that road we are simply back with the same old choice that existed long before any apologist ever thought of chaos.

If we introduce the quantum nature of reality then the game does change because the starting conditions of a complex chaotic system are now indeterminate in principle, not just in practice. Moreover, this irreducible quantum uncertainty can be amplified to a significant level on the scale of everyday experience very easily and quickly. So there do exist some aspects of things that are indeterminate and through which a deity could apparently intervene without doing violence to the invariant laws of Nature.

But there does not seem to be anything in this that has not been said by religious apologists ever since they appreciated the implications of quantum theory. There is a finite probability that jars of water will quantum-mechanically tunnel into jars of wine: that is, change between two states. No laws of Nature need be violated. But I wonder how persuasive such apologetics really are with the audience at whom they are aimed: those who understand quantum mechanics.

Moreover, as the scientists of the 18th and 19th century fully appreciated, when you consider 'random' processes, as they would have called them, it is important to explain why there exist stable long-term averages. Florence Nightingale, who in her spare time was a keen student of the fast-growing subject of statistics, regarded the emergence of such long-term stability out of a concatenation of independent random events as a hallmark of divine guidance. Polkinghorne ignores this emergence of orderly large-scale phenomena out of microscopic chaos in his discussion but it would be interesting to see him attempt to incorporate it into his view in the future.

CHAPTER 35

Accepting Pascal's wager

And how am I to face the odds
Of man's bedevilment and God's?
I, a stranger and afraid
In a world I never made.

A.E. Housman

It has been said that philosophers can be divided into two classes: those who believe that philosophers can be divided into two classes and those who don't. Brams's unusual book focuses upon the single issue most likely to split them: whether or not there exists a supreme being (affectionately referred to as 'SB' by the author).

The amount of circular discussion of this topic down through the ages is depressingly large. Whether or not Brams's approach to this question proves popular, it will not detract from the fact that he has introduced a fundamentally new idea into a study that has been pursued for millennia, and he has succeeded in shifting some problems from the realm of metaphysics into the world of mathematics, where all their consequences can be worked out systematically.

Brams, who is professor of politics at New York University and known for his work on preference voting systems, argues that it is possible to formulate various questions regarding the existence and *modus operandi* of a deity together with issues like free-will in simple mathematical terms. This mathematical attack uses game theory, a weapon that has already been successfully deployed in other studies—notably economics and evolutionary biology, where multiple choices and strategies are available. Brams asks: if there exists a superior being possessing qualities of omniscience, omnipotence, immortality and incomprehensibility, how would such a being interact with finite mortals like ourselves in situations where a variety of actions and reactions are available to both parties. These games can be formulated precisely and optimal strategies for both players

---

A review of *Superior beings: if they exist, how would we know? Game-theoretic implications of omniscience, omnipotence, immortality and incomprehensibility*, by Steven J. Brams, Springer, 1984

calculated. The idea then is to try to detect the existence of a supreme being by discovering whether observed events are consistent with the presence of such an adversary adopting a logically optimal strategy.

The type of interaction that can be analysed using game-theoretic methods is typified by Pascal's wager—that God exists or does not exist—which Brams treats in some detail. If we believe that God exists and He does, the reward is eternal happiness. If we believe that He exists and he doesn't, little is lost; and the same is the case if we disbelieve and He doesn't exist. If we disbelieve, however, that He exists and He does, we are damned for eternity. Thus, we have everything to gain and nothing to lose by wagering that God exists.

By first using two-person game theory to discover the optimum strategies of both superior being and mortal, Brams is able to determine what the stable equilibrium outcomes must be, if they exist. He goes on to consider a number of extensions of these basic tableaux. In particular, he investigates the consequences of omniscience on the part of the supreme being and demonstrates the surprising fact that omniscience can turn out to be a handicap for a player involved in a game of strategy with a mortal opponent. Another generalization from other game-theoretic applications is to allow a sequence of responses by each player to the other's moves and investigate what the long-term stable strategies must inevitably become. An amusing result of this example is that it turns out that a mixed probabilistic strategy can be optimal for the supreme being. Brams offers this as a rational explanation for a deity's apparent arbitrariness in behaviour over a long time and even the occasional injustice—a sort of mathematical existence proof of the problem of evil.

What is one to make of all this? Although it certainly makes for entertaining and stimulating reading, some of the arguments are long and complicated, primarily because these are explained in words rather than through mathematical equations. The only mathematics used explicitly in the text is some simple two-person game theory, explained as the story proceeds. Nevertheless, as with any logical analysis it is to the axioms and initial assumptions that we must look in order to evaluate the credibility of its conclusions.

One assumption that is questionable is whether mortals do always act rationally, rather than emotionally or cooperatively or mistakenly. Throughout Brams's analysis, which harks back to his analysis of interactions between Yahweh and the Jews in an earlier book on games in the Old Testament, the question of human fallibility is never examined. Also, an element of trust rather than strategy for tactical supremacy might enter, leading to a deviation from minimax options. We might wonder to what extent a mortal's choice of strategy might be influenced by other things he knows about the superior being or other things he is shown by him: is a two-person game really appropriate when decisions have to be made by an individual who interacts, perhaps tactically, with a large number of other mortals? There may also be a fallacy in treating an entire race or species as a single individual. Certainly this type of problem has arisen in what, in many ways, is an analogous methodological study: the question of whether superior extraterrestrial beings exist. In trying to decide whether or not such beings would choose to reveal themselves to us or treat us as members of a protected galactic nature reserve (such extraterrestrial superior beings are always regarded as benevolent), many have fallen into the trap of evaluating the extraterrestrials' strategy as though they were a single individual rather than a society containing a diverse spectrum of opinions.

The final chapter contains some intriguing discussion of undecidable games. These arise when, in a two-person game, the mortal player cannot decide whether the other player is superior by studying his moves. The use of the term 'undecidable' here is technically different from its other famous usage in mathematics in connexion with Gödel's theorems on the inevitable logical undecidability of some statements based on arithmetic, although its import is similar. Even if we had complete information concerning the results of an interaction between ourselves and 'something else', we might be unable to identify the existence of a supreme being by evaluating whether or not our fates turn out to be consistent with play against an omniscient optimal strategy. When the game is undecidable, none of the outcomes can be uniquely ascribed to the being's superiority.

CHAPTER 36

Getting something from nothing

Space-time. Three spatial dimensions, plus time. It knots. It freezes. The seed of the universe has come into being. Out of nothing. Out of nothing and brute geometry, laws that can't be otherwise . . . Once you've got that little seed, that little itty-bitty mustard seed— ka-boom! Big Bang is right around the corner.

John Updike

The whole research programme of trying to explain the existence of the Universe as some sort of consequence of a prior state consisting of absolutely nothing grates against one's gut feelings that 'there is no such thing as a free lunch'. Non-scientists recognise it as a fact of life that you can't manufacture something out of nothing. Scientists have underpinned this intuition and experience by discovering that the laws of Nature, which describe what changes are allowed, are equivalent to statements that some other attribute of the world always remains unchanged. The fact that the laws of Nature are the same in every direction of space and from one time to another are equivalent to the conservation of the total momentum of rotation (called angular momentum) and the conservation of total energy in any physical process. These two quantities, together with the total amount of electrical charge, are observed to be conserved in all physical processes and their status as conserved quantities is deeply entwined with the entire superstructure of physics.

If one proposes giving a scientific account of a universe coming into being out of nothing then an immediate objection seems to be that one would indeed be trying to get something out of nothing because one would have to suddenly bring into being a universe that possessed energy, angular momentum, and electric charge. This would violate the laws of Nature, which enshrine the conservation of these quantities and so the creation of the Universe out of nothing cannot be a consequence of those laws.

There is nothing wrong with this argument and it is really quite persuasive until one starts to inquire what the energy, angular momentum, and electric charge of the Universe seem to be. If the Universe possesses angular momentum then on the largest scales the expansion will possess rotation. The most distant galaxies would be moving across the sky as well as receding directly away from us. In practice, the lateral motion will be too slow for us to detect even if the Universe has a significant level of rotation. However, there are more sensitive indicators of this rotation. If we consider the effects of the Earth's rotation we see that it causes a slight flattening at the poles so that the radius of the Earth is greater at the equator than at the poles. A similar thing would occur if the Universe were rotating on its largest scales. Directions along the rotation axis would expand more slowly than others. Consequently, the microwave background radiation would be hottest if it came from the direction of the rotation axis and coolest from directions at right-angles to that. The fact that the radiation temperature is the same in every direction to a precision of one part in a hundred thousand means that if the Universe does have large-scale rotation then it must be rotating more than one trillion times more slowly than it is expanding in size. This is so small that it suggests the Universe might well have *zero* net rotation and angular momentum.

Similarly, there is no evidence that the Universe possesses any overall net charge. If any cosmic structures contain a charge imbalance between, say, the numbers of protons and electrons (which have equal but opposite electric charges) then this imbalance would have a dramatic effect since electricity is so much stronger than the force of gravity holding these structures together. In fact, it is a remarkable consequence of Einstein's theory of gravitation that if a universe is of the 'closed' sort, which will eventually contract to a future singularity, then it must have zero net charge. All the matter it contains, whether the individual elementary particles of which it is made have positive or negative charges, must total to give zero overall charge.

Finally, what about the energy of the Universe? This is the most intuitively familiar example of something that you cannot produce from nothing. But remarkably, if the Universe is 'closed' then it must also necessarily have *zero* total energy. The reason can be traced to Einstein's famous $E = mc^2$ formula which reminds us that mass and energy are interchangeable and so we should be thinking about the conservation of mass–energy together rather than energy or mass taken alone. The important point is that energy comes in positive and negative varieties. If we add up all the masses in a closed universe they contribute a large positive contribution to the total energy. But those masses also exert gravitational forces upon each other. Those forces are equivalent to negative energies that we call potential energy. If we hold a ball in our hand it has potential energy of this sort which, if the ball is dropped to the ground, will become positive energy of motion. The law of gravitation ensures that the negative energy of gravitation between the masses in the Universe must always be equal in magnitude but opposite in sense to the sum of the mc^2 energies. The total is always exactly zero.

This is a remarkable state of affairs. It seems that the three conserved quantities that prevent us getting something from nothing may well all be equal to zero. The full implications of this are not clear. But, curiously, it does appear that the conservation laws do not present a barrier to creating a universe out of nothing (or, for that matter, to having it disappear back into nothing!).

As a postscript it is amusing to note that there have even been a few speculative discussions of the possibility of creating a universe in the laboratory. The serious core of this outrageous-sounding idea derives from consideration of the local conditions that are needed to create the phenomenon of inflation in the early universe. If they were to arise naturally (or even be induced artificially) in some tiny region today then that region could begin to expand at the speed of light. Fortunately, this does not seem to be possible! In any case we note that this is not really about creating universes from nothing, only altering an aspect of the expansion of some of the local matter in our Universe—albeit in a rather dramatic fashion. This might make a good science fiction story.

To close this discussion of the scientific ramifications of creation out of nothing we should hark back to the idea that the Universe began at a singularity of space and time. This idea is no longer regarded as the most likely because of the studies that have been made of quantum cosmology. Still, one should be wary of the fact that many of the studies of quantum cosmology are motivated by a desire to avoid an initial singularity of infinite density and so they tend to focus upon the possible quantum cosmologies that avoid a singularity at the expense of those that might contain one. However, it is worth noticing that the traditional picture of the Big Bang Universe emerging from a singularity is, strictly speaking, creation out of absolutely nothing. No cause is given and no restrictions are placed upon the form of the Universe that appears, but before it does there is no time and no space and no matter. The quantum creation models hope that by pursuing a description of the Universe that explains why it develops into its present state with high probability from some inevitable quantum state we will learn why the Universe possesses so many of the unusual properties that it does.

These types of study seem to trespass on traditional theological territory. In the past, orthodox Christian theologians have been content with the overall cosmology presented by the Big Bang picture. Some have made much of the apparent beginning at a singularity that the traditional picture offered. The singularity theorems indicated that the Universe needed a beginning although it could say nothing at all about why or with what likelihood that beginning occurred. This tempted some theologians to adopt a God-of-the gaps mentality to the situation, confident that the initiation of the Universe was one gap that could never be filled by the scientific enterprise. The quantum cosmological enterprise, by contrast, appears less attractive to many of the same theologians because it attempts to describe the creation process itself using scientific laws, or to show that it is inevitable in some sense. It threatens to fill in that last gap. But those theologians who do not look to God merely as the initiator of the expansion of the Universe are unmoved by these new cosmological inquiries.

Part 10

Cosmology

The Universe is but the Thing of Things
The things but balls all going round in rings
Some of them mighty huge, some mighty tiny
All of them radiant and mighty shiny

Robert Frost, *Accidentally on Purpose*

Cosmology is in the midst of a dramatic period of new discoveries. New instruments and satellite-borne detectors continually reveal fresh and unexpected features of the Universe. Not to be outdone by the ingenuity of their experimental colleagues, theorists have been ambitiously pushing back their reconstructions of the Universe's past history towards the first instants of its expansion about 15 billion years ago.

It is a sobering thought that we have only known about the expansion of the Universe since 1929, and only in the last 10 years have we come to understand how little of the Universe's material content is luminous. Our confidence in reconstructing the Universe's distant past springs from the coming together of astronomy, the study of the largest things in the Universe, and particle physics, the study of the very smallest. At first, such a symbiosis might seem strange. But, as we look back into the Universe's past, we encounter conditions of ever-greater temperature and energy in which all forms of matter are smashed down to their smallest constituents. Conversely, by a fuller understanding of those smithereens, we have learnt that the Universe may be hiding most of its material in ghostly subatomic particles that reveal their presence only by their gravitational pull on the motions of the visible stars.

The first article in this section gives an overview of the Big Bang theory of the Universe following its prediction by Einstein's theory of gravitation, first discovered in 1915. This extraordinary mathematical theory, the most beautiful of all the structures that the human mind has uncovered, allows us to understand the overall structure of the Universe and to relate the dynamics of its expansion to the amount of matter it contains. We can reconstruct its past history and predict what fossil relics should remain today of conditions billions of years ago. Our searches have been remarkably successful. They show that we have a reliable picture of what the Universe was like just one second after its expansion began. We know this because the abundances of the lightest chemical elements in the Universe are determined by the detailed behaviour of the expansion at this early time. Remarkably, the predictions of the theory match the observed abundances. By this process of predicting what we should find in the Universe today, and searching for it with our telescopes, cosmology has become a rigorous observational science.

Despite the success of the theory of the expanding Universe in supplying a consistent description of the Universe today, puzzles remain about some of the distinctive features of the cosmic expansion. Why does it proceed at a rate that is so tantalisingly close to the great divide that separates a future of eternal expansion from one in which the expansion will eventually be reversed into contraction? Expanding universes veer steadily away from this critical divide as they expand and age. Our Universe has managed to stay very close to the divide for billions of years, so it must have begun expanding at a speed that was fantastically close to the divide at the beginning. Why? There are other puzzles of this sort. Why is the Universe so similar in the way it expands, from one direction to another, and from place to place? On the face of it, these remarkable uniformities we see in its properties—good to one part in a thousand—are very unlikely to have fallen out just by chance. Moreover, the small variations that do exist have a special pattern, one that enables them to form galaxies and clusters of galaxies with the distributions and special shapes that we see with out telescopes.

A compelling solution to these puzzles emerged in 1981. Since then, cosmologists have believed that a small change in the rate of the Universe's expansion during its first moments can solve these puzzles. A brief interlude during which the expansion is rapidly accelerated is all that is needed. This proposal is called the 'inflationary universe', theory. During the last few years, cosmologists have realised that the most critical tests of the inflationary universe theory are those that can be made by observing the patterns of variation in temperature and intensity of the heat radiation left over from the very early Universe. The inflationary theory predicts that very particular patterns should be found if a brief burst of inflated expansion did occur in our past. The instrumental sensitivity and area of sky coverage needed to test these predictions decisively requires observations from space, from far above the distorting effects of the Earth's atmosphere. The first satellite to look found fluctuations consistent with expectations but could only observe variations between points separated by more than about 10 degrees on the sky. This story is told in '*What COBE saw*'. But there is a lot more of it to come: in 2000 and 2005, two new satellites (MAP and Planck Surveyor) will be launched to look at the pattern of temperature variations over far smaller angular separations on the sky. What they see will be a decisive test of our theories about the Universe's past.

The remaining chapters in this section look back at the apparent beginning of the Universe. '*Is the Universe open or closed?*' delves deeper into the question of whether the Universe is destined to expand forever or not, and explores the problem of the dark matter that appears to pervade the Universe, in and around galaxies. '*The origin of the Universe*' reviews our developing ideas about the beginning of the Universe, culminating with the striking picture of the self-reproducing, eternally inflating Universe. This article distinguishes carefully between the 'visible universe' (that part of it from which we can receive light signals) and *the* Universe (everything there is), which may be infinite. Astronomy can only tell us about the structure and history of the visible universe. Before the introduction of the inflationary universe theory it was expected that the Universe would look roughly the same beyond our horizon of visibility as it does in the part that we can see. There was no reason to expect otherwise. The inflationary universe changed that. For the first time we have positive reasons to expect that the Universe looks quite different beyond our visible horizon.

This distinction between the entire (possibly infinite) Universe and the (finite) part that is visible to us, even with perfect instruments, is important. It means that the great cosmic questions like 'did the Universe have a beginning?' or 'is the Universe finite or infinite?' are unanswerable. Some of these considerations are developed in '*Battle of the giants*', which is a review of a book based upon a short series of lectures by Stephen Hawking and Roger Penrose in which they discuss their very different ideas about a number of key cosmological problems.

Finally, we end with some light relief. '*They came from inner space*' is a short article written at the request of an American magazine for their April Fools' Day issue some years ago. Alas, it was never published because the magazine's editors feared that it would cause widespread concern in parts of the country. It invents a slow catastrophe scenario from the terrestrial impacts of WIMPs ('weakly interacting massive particles'), discussed in '*Is the Universe open or closed?*'. They are the favoured candidates to explain the preponderance of dark matter in the Universe. Here, their properties are exaggerated only a little. A short biography of the article's imaginary author, Rolf Paoli, is attached for good measure.

CHAPTER 37

An explosion into time and space

Eddington's universe goes phut
Richard Tolman's can open and shut
Eddington's bursts without grace or tact,
But Tolman's swells and perhaps may contract.

Leonard Bacon

How, why and when did the Universe begin? How big is it? What shape is it? What is it made of? These are questions that any curious child might ask; they have also preoccupied cosmologists for decades.

Gaze into the sky on a starry night and you will see a few thousand stars, most straddling the darkness in a great swathe we call the Milky Way. This is all the ancients knew of the Universe. Gradually, as telescopes of greater and greater size and resolution have been developed, a Universe of unimagined vastness has swum into view. A multitude of stars is gathered into the islands of light we call galaxies and all around the galaxies is a cool sea of microwave radiation. This background radiation is one of the most convincing pieces of evidence that time, space and matter appear to have come about after an explosive event—the Big Bang—from which the present-day Universe emerged. It is still expanding, slowly cooling and continuously rarefying.

According to the Big Bang theory, the beginning (the Bang) was about 15 billion years ago and created a Universe that was an inferno of radiation, too hot for any atoms to survive. In the first few minutes it cooled enough for the nuclei of the lightest elements to form but only millions of years later would it be cool enough for whole atoms to appear, followed by simple molecules, and after billions of years, by the complex sequence of events that saw the condensation of material into stars and galaxies. Then, with the appearance of stable planetary environments, the complicated products of biochemistry were nurtured by processes we still do not fully understand. But how and why did this elaborate sequence of events begin? What can modern cosmologists tell us about the beginning of the Universe and how does this compare with the various creation stories of ancient times?

Creationists throughout history did not attempt to reveal anything new about the structure of the world; they aimed simply to remove the spectre of the unknown from human imaginings. By defining their place within the hierarchy of creation, the ancients could relate the world to themselves and avoid the terrible contemplation of the unknown or the unknowable. Modern scientific accounts need to achieve much more than this. The theories must be strong enough to tell us more about the Universe than what we know already. And they must also allow for predictions to be made that can be checked experimentally. They should bring coherence and unity to collections of disconnected facts. The methods employed by modern cosmologists are simple, but they are not necessarily obvious to the outsider. They begin by assuming that the laws governing the workings of the world locally, here on Earth, apply throughout the Universe—until one is forced to conclude otherwise. Typically, one finds that there are, or have been, some places in the Universe—especially in the past —where extreme conditions of density and temperature are encountered which are outside our experience on Earth. Sometimes our theories are expected to continue to work in these domains—and, indeed, do. But on other occasions we are working with approximations to the true laws of Nature, approximations that possess known limits of applicability. When we reach those limits, we must try to establish better approximations to cover the unusual new conditions we have found. Many theories make predictions that we cannot test by observation. Indeed, it is those sorts of predictions that often dictate the types of astronomical observatory or research satellite that will be required in the future. Scientists often talk about constructing 'cosmological models'. By this they mean producing simplified mathematical descriptions of the structure and past history of the Universe which capture its principal features. Just as a model aeroplane reproduces some but not all of the features of a real aeroplane, so a model universe cannot hope to incorporate every detail of the Universe's structure.

Our cosmological models are very rough and ready. They begin by treating the Universe as if it were a completely uniform sea of material. The clumping of material into stars and galaxies is ignored. Only if one is investigating specific issues, such as the origins of stars and galaxies, are deviations from uniformity considered. This strategy works remarkably well; one of the most striking features of our Universe is the way in which the visible part of it is so well described by this simple picture of a uniform distribution of material. An important feature of cosmological models is that they must be open to being tested against the real Universe. Cosmologists do this by incorporating properties such as density or temperature—which can be found only by measurement—into the model. If the model is robust and a good representation of what actually happens in reality, then the properties as given by the model should match those as measured in reality.

Our exploration of the Universe has taken off in different directions. Besides satellites, spacecraft and telescopes, we have employed microscopes, atom-smashers and accelerators, computers and human thinking to enlarge our understanding of the entire cosmic environment. Besides the world of outer space—the stars, galaxies and great cosmic structures—we have come to appreciate the labyrinthine subtlety within the depths of inner space. There we find the subatomic world of the nucleus and its parts: the basic building blocks of matter—so few in number, so simple in structure, but in combination capable of being organised into the vast panoply of complexity we see around us and of which we are a part. These two frontiers of our understanding—the small world of the elementary parts of matter and the astronomical world of the stars and galaxies—have come together in unexpected ways in recent times. Where once they were the domains of different groups of scientists attempting to answer quite different questions by separate means, now their interests and methods are intimately entwined. The secret of how galaxies came into being may well be fathomed by the study of the most elementary particles of matter in particle detectors, buried deep underground. Similarly, the identity of those elementary particles may be revealed by observations of distant starlight. And as we cosmologists try to reconstruct the history of the Universe, searching for the fossil remnants of its youth and adolescence, we find that we are beginning to explore the same territory as nuclear physicists looking at the smallest pieces of our physical world with their atom-smashers. Our appreciation of the unity of the Universe becomes more impressive and more complete.

When Albert Einstein published his general theory of relativity in 1915 there was no widespread belief that the Universe was populated by those huge collections of stars we know as galaxies. It was commonly held that these extra-terrestrial sources of light—or 'nebulae', as they are called—lay within our own Milky Way galaxy. Nor had there ever been any proposal by astronomers or philosophers that the starry Universe was anything but static. It was into this intellectual ambience that Einstein launched his new theory of gravitation. Unlike Isaac Newton's classical description of gravitational forces, which Einstein's theory included and superseded, the general theory of relativity had the extraordinary ability to describe entire universes, even if they were infinite in extent.

When Einstein began to explore what his new equations revealed about the Universe, he set about doing what scientists generally do—simplifying the problem to be solved. He assumed, for instance, that the Universe looked the same in every direction, and we now know this to be an excellent approximation. But what Einstein discovered to his great chagrin was that his equations demanded that universes of this sort either expanded or contracted with the passage of time. Einstein was deeply troubled by this prediction of his theory, which he thought couldn't be true. He set about looking at the mathematics and saw that it allowed him to introduce a new repulsive force into his theory of relativity. He found a model where the repulsion exactly counterbalanced the attraction of gravity to produce a static Universe. In fact, as he was later to recall, Einstein had made what was probably the biggest blunder of his life. There was no need to introduce the repulsive force because as Edwin Hubble, an American astronomer, discovered in the 1920s, the Universe really is expanding. Hubble recognised that the light from distant galaxies displayed a phenomenon well known in physics where an object travelling away from an observer appears slightly redder than when stationary—called redshift. What Hubble had discovered was the expansion of the Universe. Instead of a changeless arena in which we could follow the local perambulations of planets and stars, he found that the Universe was in a dynamic state. This was the greatest discovery of 20th-century science, and it confirmed what Einstein's general theory of relativity had predicted about the Universe: it cannot be static. The gravitational attraction between galaxies would bring them together if they were not rushing away from each other. The Universe can't stand still.

If the Universe is expanding, then when we reverse the direction of history and look into the past we should find evidence that it emerged from a smaller, denser state—a state that appears once to have had a zero size. It is this apparent beginning that has become known as the Big Bang.

There are important things to appreciate about the expansion of the Universe. First of all, what exactly is expanding? In the movie *Annie Hall*, Woody Allen is discovered on his analyst's couch telling of his anxiety about the expansion of the Universe: 'Surely this means that Brooklyn is expanding, I'm expanding, you're expanding, we're all expanding.' Thankfully, he was wrong. *We* are not expanding. Nor Brooklyn. Nor the Earth. Nor the solar system. Nor, in fact, the Milky Way—our galaxy. Nor even those aggregates of thousands of galaxies we call 'galaxy clusters'. These collections of matter are all bound together by chemical and gravitational forces between their constituents—forces that are stronger than the force of expansion. It is only when we get beyond the scale of great clusters of hundreds and thousands of galaxies that we see the expansion winning out over the local pull of gravity. For example, our near neighbour the Andromeda galaxy is moving towards us because the gravitational attraction between Andromeda and the Milky Way is larger than the effect of the universal expansion. It is the galaxy clusters, not the galaxies themselves, that act as the markers of the cosmic expansion.

A simple picture might be to think of specks of dust on the surface of an inflating balloon. The balloon will expand and the dust specks will move further apart, but the individual dust specks will not themselves expand in the same way. They will act as markers of the amount of stretching of the rubber that has occurred. Similarly, it is best to think of the expansion of the Universe as the expansion of the space between clusters of galaxies.

What is intriguing about the concept of an expanding Universe is that it must have expanded from something smaller. The term 'Big Bang' was actually coined to explain this by the astronomer Fred Hoyle; he used it in 1950 during a BBC radio broadcast as a perjorative description of a cosmology in which the Universe expanded into being from a dense state at a finite time in the past. (Hoyle himself was sceptical about this idea and an arch-advocate of his own 'steady-state' theory of a constant density Universe.) The dispute between the Big-Bang theorists and the steady-state adherents was finally resolved in 1965 when the American astronomers Arno Penzias and Robert Wilson discovered something called microwave background radiation. Such heat radiation—found to be everywhere in the Universe where radio astronomers looked—would not be present in a steady-state Universe because such a Universe would not have experienced a hot past of enormous density; rather, it would have been, on average, rather cool and quiescent at all times.

The discovery of cosmic microwave background marked the beginning of serious study of the Big Bang model. Gradually, other observations revealed further properties of background radiation. It had the same intensity, or 'brightness', in every direction to at least one part in a thousand. And, as its intensity was measured at different frequencies, it began to reveal characteristic variations of intensity with frequency—the signature of pure heat. Such radiation is called 'black-body' radiation. Unfortunately, the absorption and emission of radiation by molecules in the Earth's atmosphere prevented astronomers from confirming that the whole spectrum of the radiation was indeed that of heat radiation. Suspicions remained that it might have been produced by violent events that occurred nearby in the Universe long after the expansion had begun, rather than by the Big Bang itself. These doubts could only be overcome by observing the radiation from above and beyond the Earth's atmosphere. This was finally done by Nasa's Cosmic Background Explorer (COBE) satellite in 1992. The instruments on board the satellite measured the most perfect black-body spectrum ever seen in Nature. It was striking confirmation that the Universe was once hundreds of thousands of degrees hotter than it is today. Another key experiment to confirm that the background radiation did not have a recent origin nearby in the Universe was carried out by high-flying U2 aircraft. These former spyplanes are extremely small with large wing spans, which makes them very stable platforms for making observations. On this occasion, they were looking up rather than down, and they detected a small but systematic variation in the intensity of the radiation around the sky, a variation that had been predicted to arise if the radiation originated in the distant past.

The microwave background radiation is like an ocean through which all heavenly bodies are moving. Each is moving within its own system—for example the Earth and planets around the Sun—and each system is moving within another—for example the Sun's motion around the Milky Way and the Milky Way's motion around the Sun, and so on. This means that we are moving through radiation in some direction. The radiation intensity will appear greatest when we look in that direction and least intense 180 degrees away, and should display a characteristic variation in between. It is rather like running in a rainstorm. You get wettest on your chest and stay driest on your back. This is precisely the sort of variation in microwave intensity that the U2s detected.

The microwave background radiation found everywhere in the Universe fitted the Big Bang theory perfectly. Moreover, subsequent observations of the abundance of the lightest elements in the Universe matched the predictions of the Big Bang model and confirmed the idea that they were produced by nuclear reactions during the first three minutes of the expansions. The steady-state model offers no explanation for these abundances, because it never experiences an early period of great density and temperature when nuclear reactions can occur throughout the Universe. These successes were the death knell for the steady-state model, and it played no further role as a viable model for the Universe, despite attempts by some of its advocates—Hoyle in particular—to modify it in various ways. The Big Bang model has established itself as the one that succeeds in co-ordinating our observations of the Universe. But it is important to understand that the term 'Big Bang model' has come to mean nothing more than a picture of an expanding Universe in which the past was hotter and denser than the present. The mission of cosmologists in the future will be to pin down more precisely the expansion history of the Universe and to determine how galaxies formed. They will be asking why galaxies cluster as they do, why the expansion proceeds at the rate it does—and trying to explain the shape of the Universe and the balance of matter and radiation existing within it.

CHAPTER 38

What COBE saw

Telescope, n. *A device having a relation to the eye similar to that of the telephone to the ear, enabling distant objects to plague us with a multitude of needless details. Luckily it is unprovided with a bell summoning us to the sacrifice.*

Ambrose Bierce

Our eyes are sensitive only to a narrow range of the wavelengths of light. This 'visible' part of the spectrum of light spans the colours of the rainbow from red to violet and once provided our only vista on the Universe. Light beyond this small band of frequencies is either insufficiently energetic to register a signal on the chemical sensors of the eyes or so energetic that it would damage them. Modern advances in astronomy can be ascribed to the development of artificial eyes that detect and amplify cosmic background light across the whole range of the spectrum. Some parts, like the X-ray and ultraviolet bands, hardly penetrate the Earth's atmosphere. Others, like the microwave band, penetrate the Earth's atmosphere but are slightly distorted by interference from interactions with atoms in the atmosphere. To study the appearance of the Universe in wavebands which are distorted by atmospheric interference, we need to deploy observatories in space. At first these were balloons, brief rocket flights or high-flying aircraft, but during the last decade they have escalated to include satellite-borne observatories. The Cosmic Background Explorer (COBE) satellite carried a suite of instruments designed to study, with unprecedented precision, the background of microwaves that permeate our Universe.

Although the 'visible' part of the spectrum dominates our impressions of the Universe, and is the cornerstone of astronomy, most of the light in the Universe lies in the microwave frequency range. This sea of cosmic microwaves carries with it the most potent pieces of information about the past history of the Universe. Its properties are extraordinary. The microwave light contains more energy than has been produced by all the stars in the Universe put together and its intensity is identical in every direction of the sky to better than one part in a thousand. No population of cosmic objects could conspire to produce such fidelity when their signals were combined. This radiation is not local. It cannot originate in stars and galaxies. It comes from the beginnings of the Universe, an echo of the Big Bang.

The expansion of the Universe, predicted by Einstein's theory of general relativity, was discovered at the end of the 1920s by the American astronomer Edwin Hubble. If, in our mind's eye, we reverse that expansion, we can envisage a past in which the Universe was hotter and denser than it is today. Eventually, about fifteen billion years ago, our reconstruction encounters an apparent beginning to the expansion. The microwave radiation we see today is the radiation fall-out from the Universe's first moments. As it expanded and cooled, allowing first nuclei, then atoms and molecules, then planets and people, to form, so the radiation temperature fell steadily to its current value of just 2.73 degrees Kelvin. As early as 1948 two young American physicists, Robert Herman and Ralph Alpher had predicted that radiation fall-out should remain from the Big Bang with a temperature close to this value. Unfortunately, their prediction never became widely known and the radiation was discovered serendipitously at Bell Labs in New Jersey in 1965 without its discoverers realising what it was. Its significance was soon pointed out by others and its detailed study has become the observational cornerstone of modern cosmology.

The COBE satellite has made two dramatic discoveries. The first concerns the detailed form of the spectrum of the radiation: the variation of its intensity with frequency across the whole microwave range. Complete information cannot be obtained with accuracy from the ground because of the distortions produced by the Earth's atmosphere. It was long-awaited by cosmologists because the Big Bang origin of the radiation leads one to expect that it will display the characteristic spectrum of heat radiation but perhaps with little blips superimposed here and there if violent events occurred in the past when galaxies and stars formed. Ground-based observations recorded distortions but were they fossils of primeval history or merely noise from the atmosphere? In 1987 all was revealed. When the COBE satellite's observations were made they revealed the most perfect spectrum of heat radiation ever observed in Nature. No distortions were present down to a limiting accuracy better than one part in a hundred. The unblemished spectrum tells us that the early history of the Universe was quiescent and eliminates from consideration an array of violent scenarios for the origin of galaxies.

COBE's second great discovery was to measure the first variations in the intensity of the microwaves from one direction in the sky to another. If the Universe is expanding slightly faster in one direction than in another then the radiation intensity from the first direction will be slightly reduced with respect to the second. Likewise, galaxies arise from regions of the Universe with slightly more material in them than the average. When the radiation passed through those regions it would cool differently than if it had passed them by. A map of how the microwave intensity varies around the sky gives a picture of how the gravity field of the Universe looked in the distant past; it helps us to determine the sizes of the precursors of galaxies and clusters of galaxies and to discover whether the overall expansion of the Universe is asymmetrical in any way. It would also provide a further confirmation of the whole picture of the expanding Universe in which galaxies begin as regions of greater-than-average density that eventually separated from the overall expansion to form great islands of stars. We see them today and so they should have existed in the past and left their gravitational imprints upon the radiation around them.

COBE was the first observatory to see these tiny microwave fluctuations: they are among the smallest effects measured in astronomy—variations in radiation intensity of just a few parts in one hundred thousand. Finally, after more than a quarter of a century of searching, the seeds of the structures that illuminate the present Universe have been found.

To appreciate the significance of the COBE measurements it is helpful to consider an analogy. Suppose one knew a fair amount of biochemistry and genetics but was observing a human being for the first time. If this person was an adult you might begin to speculate about how rapidly it could have grown. All this speculation continues for a long time with many rival ideas proposed about human growth. Then one day someone sees a baby. This observation allows you to check your theories about how small humans were in infancy; how fast they must grow to reach maturity in a given time and how their shapes evolve during that growth process. The COBE satellite has done something similar. It has taken a photograph in radio waves of the Universe when it was about a million years old. Today it is about fifteen billion years old. COBE's microwave photograph shows the Universe before galaxies formed when their embryos were merely blips on the cosmic landscape.

Theories about the nature of the Universe during the first fleeting moments of its expansion predict that particular collections of irregularities will be created then and will eventually leave their signatures in the microwave background radiation. COBE's observations allow these theories to be tested directly. They offer new insights into many of the problems that remain unsolved. Other observations of the complicated clustering patterns of the galaxies on the sky tell us about the later history of galaxy evolution. Are those patterns consistent with the COBE observations if they are both interpreted in the light of our present theories of galaxy formation? Can we recreate the essentials of the entire process of galaxy formation and clustering by computer simulations? Can we understand why some regions of the Universe turn into featureless non-rotating swarms of stars whilst others become swirling spiral discs like the

Milky Way? And why do the stars and galaxies all move as if responding to the pull of a gravity field that is ten times greater than could be created by all the matter we can see in the Universe? What is the nature of that mysterious hidden matter that is doing the pulling? Why is it dark and is there enough of it out there one day to draw the expansion of the Universe to a close and bring the skies down upon our heads? Our best chance of answering questions like these is by flying further satellite missions which can see finer structures than were visible to COBE. In 2000 and 2005 two new missions will be flown by NASA and the European Space agency which will be able to probe the fine structure of the microwave sky at a level that will tell us how the galaxies formed and test our theories about the early history of the Universe.

CHAPTER 39

Is the Universe open or closed?

That's the problem with eternity, there's no telling when it will end.

Tom Stoppard

Until only 75 years ago it was the opinion of all scientists and philosophers who had ever thought about the matter that space was an unchanging background stage upon which the motions of the heavenly bodies were acted out. It was absolute, unchanging, unbending, never-ending: the laws of Nature provided the script for all the motions played out upon it.

This picture of the Universe of space was endorsed by Isaac Newton's laws of motion and gravitation. The laws assumed the pre-existence of a space within which things move. That space was said to be 'flat'. It was the space that obeys Euclid's famous geometry. If one picks any three points on a flat surface and then joins the first to the second, the second to the third and the third to the first by the shortest possible paths, then one has drawn a triangle with three straight lines for sides and the sum of the three angles at the inside corners of the triangle is always 180 degrees.

In 1915 Albert Einstein changed our view of space in a radical way. He gave up the view of space as a fixed billiard table, replacing it by a theory in which space was akin to a rubber sheet that would be deformed by the motions of objects upon it. No longer was space something separate from matter and motion: the presence and motion of matter determined the geometry of space, whilst the geometry of space dictated how matter would move upon it. In places where there was no matter, space would be flat with a geometry of Euclid's sort; but in places where there was a concentration of mass the geometry would be deformed and curved so that if one were again to draw paths of least distance around a trio of points one would have a curved triangle whose internal angles would not total 180 degrees. This is just as it would be if we flew around a circuit of least distances joining three cities on the Earth's surface. If the curvature is like that on the surface of a sphere then the sum is greater than 180 degrees but if the surface has an opposite sense like that on a saddle then the sum will be less than 180 degrees. Einstein's revolutionary theory of gravitation—the general theory of relativity—showed how one could calculate the geometry of space in the presence of any quantity of matter, moving in any manner at all.

By the end of 1920s a number of mathematicians had discovered that this new way of looking at gravitation explained several puzzling observations perfectly and also allowed them to produce mathematical descriptions of the entire Universe. These descriptions of the Universe all had the remarkable feature of predicting that the Universe should be expanding with time, a prediction that was dramatically confirmed by Edwin Hubble's observations of the systematic redshifting of the light from distant galaxies. The distant clusters of galaxies which contain these galaxies of stars are all receding away from each other rather like raisins in a ball of rising dough being heated in an oven. But these expanding universes could be of three types. The first is called a *closed universe*. In such a world the energy of the expansion is overwhelmed by the deceleration created by the gravitational pull of all the matter in the Universe and eventually the expansion will be halted and reversed into contraction. It is like throwing a stone in the air; the gravitational pull of the Earth prevents it escaping from the Earth's surface. The second is called an *open universe* and here the expansion energy wins out over the pull of gravity and the expansion will continue for ever. It is like launching a rocket from the Earth at a speed exceeding the 'escape velocity' of 11 kilometres per second. It will not return to Earth. The third variety of universe (which I call the 'British compromise universe') is on the critical dividing line between the 'closed' and 'open' universes. The expansion energy exactly balances the pull of gravity and this 'critical' universe just manages to expand forever. These are illustrated in Figure 39.1.

Why are these universes called 'open' and 'closed'? We explained how the presence of material in space determines its shape by analogy with objects on a rubber sheet. This applies to the Universe as a whole. All the matter in the universe determines the overall shape of the space in which it sits. The critical universe has precisely that density of matter within it that is counter-balanced by its expansion so as to allow space to be flat like Euclid's. But, when the density of material in the universe is greater, the space is curved up like that on a closed surface and we have a closed universe. By contrast, when the density is less than the critical value the curvature has the opposite sense and resembles the curvature on the surface of a saddle, see Figure 39.2.

How does this connect with the real world? Einstein's equations that connect the curvature of space to the material within it enable us to calculate the critical density that separates the ever-expanding 'open' universes from the 'closed' universes with their finite future histories. It depends upon the rate at which the Universe is expanding. It corresponds to the very small density; if we were to smooth out all the material in the stars and galaxies into a uniform sea of

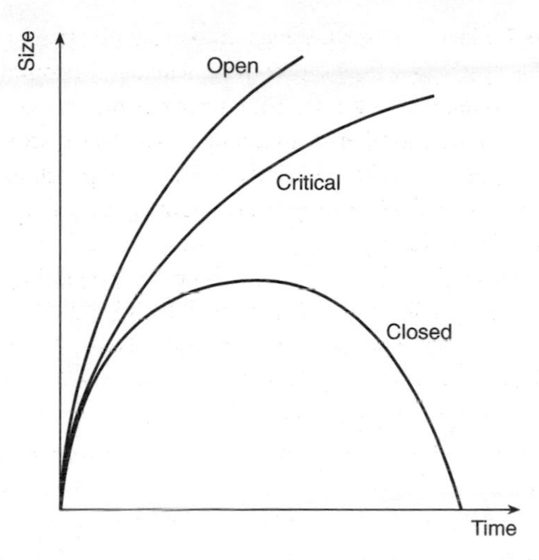

Fig. 39.1. The three varieties of expanding universe. 'Open' universes are infinite in extent and expand forever. 'Closed' universes are finite and contract to a 'big crunch'. The divide between the two is marked by the 'critical' universe, which is infinitely large and expands forever.

atoms then the critical density that must be exceeded in order to 'close' the Universe is only about 10 atoms in every cubic metre. This is minute—far more rarified than any man-made vacuum produced in a laboratory on Earth. Of course, in the actual Universe, material is aggregated in concentrations and lumps, like galaxies, stars, planets, and even people. Most of the things we see around us have a density about 10^{29} times greater than the critical average density. This shows us that outer space is almost entirely simply that—empty space.

So, how dense is *our* expanding universe? Is there enough material within it to exceed the critical value and close the universe or does it fall short and leave us facing an agoraphobic future of infinite expansion? The answer creates some of the greatest problems of cosmology. If we total all the material that we can detect in the Universe—the luminous stars and galaxies, the dust and grains of matter, the radiation and cosmic rays—we find only about 10 per cent of the critical density, but this does not allow us to claim immediately that the Universe is open and destined to expand forever. For, if we observe the motions of stars in galaxies and of galaxies in clusters of galaxies something very startling emerges. These objects are found to be moving so fast that the gravitational pull of all the luminous forms of matter around them could not stop those stars and galaxies from escaping from their respective galaxies and clusters. If what we can see is all there is of the Universe then all galaxies and clusters of galaxies should have disintegrated billions of years ago.

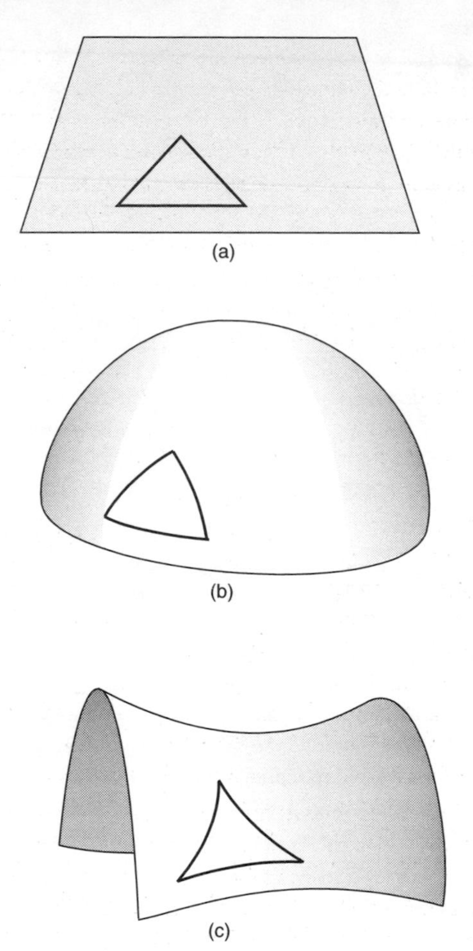

Fig. 39.2. (a) A Euclidean triangle on a flat surface and two non-Euclidean triangles on (b) closed and (c) open curved surfaces. On the flat surface the interior angles of the triangle sum to 180 degrees, on the open surface to less than 180 degrees. On each surface the definition of a 'straight line' joining two points is the shortest distance between them that lies on the surface.

What has been concluded from this state of affairs is that the Universe contains lots of dark material intermingled with the stars and galaxies. It is the gravitational effect of this dark matter that maintains galaxies and clusters in a stable equilibrium. Nor is this conclusion unreasonable. It is rather likely that star formation and galaxy formation is not 100 per cent efficient and lots of material gets left behind in these processes. Moreover, luminous stars and galaxies will only form at places where the density of matter is above some threshold. Places where the density is very low will not initiate star formation and material will be left unincorporated into luminous stars.

How much of this dark matter is there? The simple answer is that we do not know. If there was enough to stabilize galaxies and clusters then we might still get away with having a total density for the Universe of only about 20 per cent of that required for a closed universe. But there is no reason why there should not be more dark material than this. There could easily be more unseen material than is necessary to close the Universe. Only if we try to accommodate more than about five times the critical density in unseen forms do we run into conflict with other observations—it would slow the expansion so much that the expanding Universe would have lasted for so short a time that the time since the expansion began would be less than the age of the oldest rocks on Earth!

So, the cosmologist cannot say for sure whether the Universe is 'open' or 'closed' because the issue can only be decided when we know how much dark matter there is. If we find less than the critical density we might always be failing to find all the material, so we could never be sure the Universe was 'open'. On the other hand, if we track down more than the critical density we would be able to conclude that the Universe was 'closed'. But the major issue in modern cosmology is to ascertain the identity of this dark matter: is it in the form of rocks, faint stars, dust, atoms, gas, or some entirely different, exotic, form of matter?

One of the great successes of the Big Bang theory of the expanding universe is that it predicts that during the first three minutes of the Universe's expansion conditions would have been hot enough and dense enough to initiate nuclear reactions throughout the Universe. These reactions produce the abundances of the lightest nuclear elements of hydrogen, deuterium, the two isotopes of helium, and lithium. Remarkably, the predicted abundances of all these elements match those observed everywhere we look if the density of material participating in nuclear reactions corresponds to the value of the density we observe in all the luminous material in the Universe after we have allowed for the expansion of the Universe between then and now. This tells us two very important things: first, all the ordinary matter of the sort that takes place in nuclear reactions can be accounted for by the material we can see with our telescopes; secondly, in order to explain the stability of galaxies and clusters we need much more dark material and this material must be in some form that does *not* take part in nuclear reactions during the early stages of the Universe.

There are natural candidates for this special type of dark material. One class of elementary particles of matter, called *neutrinos*, do not participate in nuclear reactions. They do not carry any electric charge so they do not participate in electromagnetic interactions either. They interact only through the two weakest forces in Nature, gravity and the so called 'weak force' that is responsible for radioactivity.

The contribution that these ghostly particles make to the Universe depends upon how heavy they are. We know of three different varieties of neutrino and none of them has ever been found to possess a mass different from zero. But such evidence is not taken very seriously; because they interact so weakly these experiments are very difficult and they would not be able to detect the neutrino's mass if it is rather small.

But particle physicists have more to offer us. Their theories, which strive for the unification of all the forces of Nature, predict the existence of many other weakly interacting, neutrino-like, particles called WIMPs (an acronym for weakly interacting massive particles) that we could not have detected in the laboratory as yet. One of the aims of the new generation of particle accelerators and colliders in Geneva and the US is to discover some of these particles by creating a high-energy environment like the early stages of the Universe in which they are profusely created.

If the known neutrinos possess masses that add up to no more than about 90 electron volts (an atom of hydrogen has a mass of about a billion electron volts) then all of those neutrinos throughout the Universe will contribute a density that exceeds the critical value and the Universe will be closed. Alternatively, if WIMPs exist and have masses just two or three times that of the hydrogen atom then their cumulative density will exceed that required to close the Universe.

If the Universe is primarily composed of a sea of these weakly interacting particles then we might ask why we cannot detect them directly and settle the matter once and for all. Unfortunately, if the dark matter is in the form of the known neutrinos, with their tiny masses, we have no hope of ever detecting them directly. They interact too feebly with our detectors. All we can do is attempt to measure the masses of the neutrinos in the laboratory and then test our expectations of what these particles will do to the clustering of luminous matter by comparing computer simulations of the clustering process with observations. In fact, such investigations indicate that the clustering of galaxies that would be produced is rather unlike what is seen in the sky. However, if it is the WIMPs that constitute the dark matter then things are much more exciting. They seem to produce the right type of clustering of luminous galaxies according to our computer simulations. Moreover, these particles are a billion times more massive than the known neutrinos could be and they hit our detectors much harder. In fact, it is within our capabilities to detect a sea of these particles in the Universe around us if they do indeed constitute the dark matter. And, at present, several experimental groups in Britain, Europe, and the USA are constructing underground detectors hoping to discover the cosmic sea of WIMPs. When one of these particles hits an atomic nucleus in a crystal it will cause it to recoil and register a signal by slightly heating the crystal with the energy deposited. If one monitors just a kilogram of material for these characteristic events one should find between about one and 10 events every day. If one can shield out all the other particles, from cosmic rays, radioactive decays, and other terrestrial events, that would otherwise swamp the detector, then it should be possible to determine whether WIMPs are indeed all around us. This shielding is achieved by siting the detector deep underground, locating it within a fridge that cools to within a few hundredths of a degree of absolute zero (−273°C) (to minimise the thermal agitation from the local environment and exploit the remarkable response of low-temperature materials to being hit by WIMPs), and surrounding it with absorbent materials and sensing devices.

Over the next few years we will begin to see the results from experiments like this. They promise to reveal remarkable things about the Universe in unexpected ways. The issue of whether the Universe is 'open' or 'closed' may hinge upon the properties of the smallest particles of matter and may be decided at the bottom of mineshafts on Earth rather than with telescopes or satellites. The great clusters of galaxies may be but a drop in the ocean of matter in the Universe. The bulk of that matter, perhaps enough to curve space up into closure, may reside in a form of matter quite unlike that which we have yet detected in our particle accelerators. This could be the final Copernican twist to our status in a material Universe. Not only are we not at the centre of the Universe; we are not even made out of the predominant form of matter in the Universe.

CHAPTER 40

The origin of the Universe

The Universe is not on the side of frugality.

Don Marquis

There has been speculation about the source and nature of the Universe in most human cultures throughout most of recorded history. These speculations have involved relatively few real alternatives: either, that the Universe had no origin and will have no end, or that it had its beginning at a finite time in the past, or that what we see and experience is a transient phase or ordered reality in a chaotic Universe, or that an exploration of the past will lead us ultimately to a reality that transcends commonplace notions like time and space. These alternatives are all still very much with us, albeit in the guise of sophisticated mathematical theories of the Universe.

The first significant applications of physics to the structure of the Universe that led to strong arguments for an origin in time arose in the 19th century. They were consequences of the development of the laws of thermodynamics which emerged from the Victorian fascination with machines and industry. In a climate of industrial revolution it was easy to think of the Universe as a great engine whose workings were subject to the second law of thermodynamics, which ordains that the entropy (in some precise sense a measure of the disorder) of a closed system can never decrease. The application of this law to the Universe as a whole, in 1850, led Rudolf Clausius to the now familiar notion of the 'heat death of the Universe'. Later, others, notably the English philosopher of science William Jevons, in 1873, imagined running the consequences of the second law backwards in time to conclude that 'there was an initial distribution or violation of natural laws at some past date'. Thus, by tracing the present disordered state of the Universe backwards in time he concluded that the second law required that a state of zero (or minimal entropy) must be reached at a finite time in the past which qualified as the start of the Universe or at least the start of the present regime of natural laws.

Fifty years later, with the advent of the first models of the expanding universe based on Einstein's theory of general relativity, the same argument was framed by astronomers like Arthur Eddington in his popular works, to argue for a beginning to the expanding universe. Incidentally, we can see that these arguments could have been criticised at the time of their inception. A quantity that is always increasing does not need to have been zero at any finite past time (consider, for example, entropy increasing exponentially with the passage of time).

The discovery by Alexander Friedmann, in 1922, of the first solutions of Einstein's field equations which describe expanding universes was followed by a long period of uncertainty and debate about the meaning of the redshifting of light from distant galaxies (or 'nebulae' as they were then referred to) culminating in Hubble's deduction of the expansion of the Universe by interpreting redshifts as Doppler shifts from receding sources of light. Two coarse-grained alternatives were presented by Friedmann's solutions: universes that expand forever and universes that will reach a maximum size before contracting (see Figure 39.1, p. 242). In all cases the solutions predicted a state of infinite density (a 'singularity') at a finite proper time in the past. This invited interpretation as the 'beginning' of the Universe, although it was always possible to suspend the mathematics momentarily at the singularity and continue through it to reach the contracting phase of a previous cycle of expansion and contraction, as shown in Figure 40.1.

Modern astronomy has confirmed the expansion of the visible universe in some detail and revealed that we live about 15 billion years after the expansion of the Universe would be judged to have begun, if we assume that Einstein's equations hold all the way back to the beginning and that we know all the aspects of physics that might have a bearing on the behaviour of the Universe at very high density. In fact, we are only able to produce a good model of the early

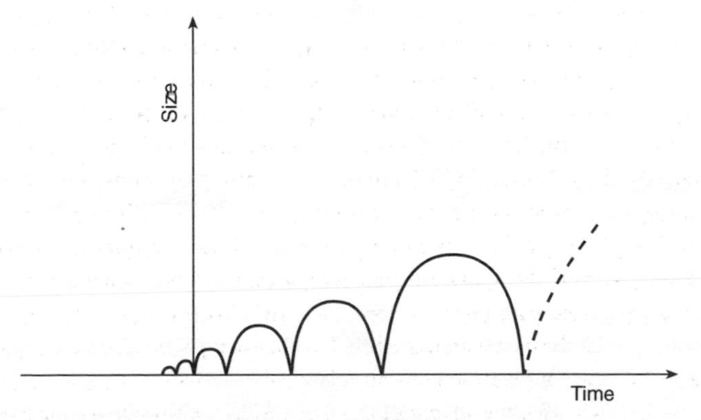

Fig. 40.1. A cyclic universe in which a closed universe 'bounces' through a succession of cycles. It is believed that the increase of entropy required by the second law of thermodynamics demands that the oscillations increase in size, as shown.

universe back to a time that lies one second after the apparent beginning. Before this time we need a fuller description of the behaviour and identity of elementary particles of matter than we possess at present. When the Universe is younger than about 10^{-10} seconds, it experiences extremes of temperature and energy that exceed those we can produce on Earth in particle colliders such as that at CERN and elsewhere. However, it should be noted that when the Universe is between one second and three minutes old it behaves like a vast nuclear reactor producing definite abundances of the light isotopes deuterium, helium-3, helium-4, and lithium-7, which correspond to the abundances observed today. Those detailed astronomical observations allow us to confirm the consistency of our model of the Universe right back to these very early times. Likewise, the existence of the microwave background radiation—its thermal spectrum, dipole temperature variation, and flat spectrum of higher multipole temperature fluctuations (all observed with great precision by the COBE satellite and by ground-based and airborne observations) confirm the general picture of the hot early universe.[1] This is usually referred to as the 'Big Bang' model and is accepted as the working picture for the evolution of the Universe by almost all cosmologists. Many aspects of its evolution (even at very late stages) are still unknown. There are many detailed variants of the Big Bang model, especially with regard to the origin of galaxies. Here we are going to focus attention entirely upon modern ideas about the apparent beginning to the expansion of the Big Bang Universe models rather than upon the details of their evolution and recent history.

First, we should say that our observational evidence does not compel us to conclude that the expanding universe had a beginning in a state of infinite density. The observed universe is compatible with all sorts of other behaviours as the apparent beginning is approached, none of which involve encountering a singularity. A few examples are shown in Figure 40.2.

In the 1930s there was already some debate about the reality of the apparent singularity in the cosmological models of expanding universes. This debate continued without a clear resolution until 1965. Three doubts were raised about the physical reality of the singularity at the beginning of the expanding universes.[2] The first was that it might be avoided if pressure was included in the model of the Universe. Just as a squeezed balloon resists compression because of gas pressure, so complete compression of the Universe might be impossible because of the radiation and gas pressure exerted by its constituents. Unfortunately, when pressure was included in the cosmological models the singularity remained (the reason is quite subtle; in essence it is because pressure is also a source of gravitation in general relativity and its inclusion actually hastens the appearance of the singularity in the past).

The next suggestion was that the singularity appeared because our models of

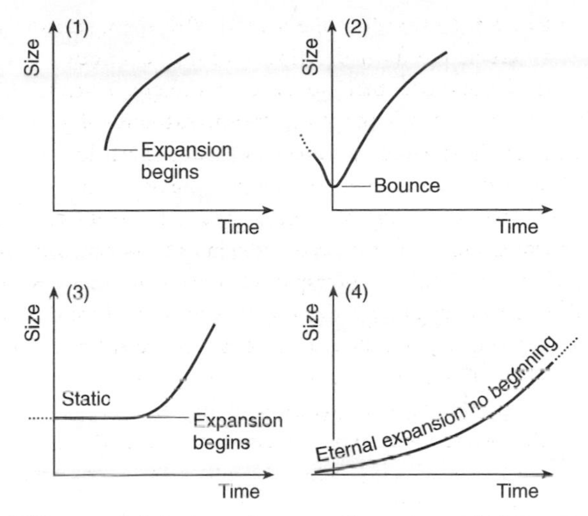

Fig. 40.2. Different early behaviours of an expanding universe which are compatible with astronomical observations of its present-day behaviour. These unusual behaviours arise from adding unusual forms of matter or generalising Einstein's equations in the high density environment near the apparent beginning.

the Universe were artificially symmetrical: that is, they expanded at the same rate in every direction and so when run backwards all the matter appeared to pile up at one point at one time. Cosmologies were then investigated in which the Universe expands at different rates in different places, but they were all found to possess an initial singularity.

The last doubt was more subtle. It was suggested that the singularity might be merely a singularity of the coordinates used to describe the expanding universe rather than of the Universe itself. Just as the meridians intersect on approach to the poles on a geographer's globe (even though nothing odd happens to the Earth's surface at the poles) so we might be employing a bad set of map coordinates for the Universe. This possibility was difficult to exclude completely. Every set of coordinates that was tried hit a singularity and although quantities, like the density of matter, that do not depend on the coordinates became infinite there, it could always be argued that these conclusions were just artefacts of looking at a very small number of special cosmological solutions of Einstein's equations. You needed to have a general solution of Einstein's equations (without any symmetries at all) to be sure, but we are unable to find a general solution in closed form (the equations are probably not soluble, in fact, because they exhibit chaotic behaviour in general). Every solution that was examined did possess a singularity. Unfortunately, an error of logic by two

Russian physicists led them to draw the erroneous conclusion that the general solution would not contain a physical singularity because all the solutions that had been found to contain singularities were solutions with special symmetries.

All these uncertainties were removed in 1965 when Roger Penrose found a new way to attack the problem. Instead of seeking particular solutions of Einstein's equations, he introduced powerful mathematical techniques from differential topology to investigate the properties of the set of all past histories of light rays and massive particles in the universe.[3] Irrespective of special symmetries or the choice of coordinates he was able to display simple conditions under which it was inevitable that at least one of these histories had a beginning in time (that is, could not be extended indefinitely into the past). The first singularity theorem of this sort was proved by Penrose for the gravitational collapse of a black hole. In collaboration with Stephen Hawking in 1970 he proved a version of the theorem that applied to the Universe.[4,5]

The Hawking–Penrose theorem states that *if*: (1) space–time is sufficiently smooth; (2) time travel is impossible; (3) there is enough matter and radiation in the Universe; (4) general relativity (or some similar theory of gravitation) holds; and (5) gravity is always attractive, i.e.

$$\rho + 3p/c^2 > 0 \qquad (*)$$

(where ρ is the mass density and p is the pressure of matter in the Universe, and c is the velocity of light), *then* there is at least one causal history with a beginning in the past; that is, *a singularity*.

There are a number of things to notice about this result. First, it is a *theorem* not a *theory*. The conclusion can only be rejected by rejecting one of the assumptions [although many variants of this theorem have subsequently been proved which weaken the requirements for a singularity or replace some of the assumptions (e.g. condition (2)) by others]. The theorem does not require every light ray or particle history to have a beginning, nor is the singular beginning necessarily accompanied by infinites in the density of matter or the curvature of space–time, although there have been many attempts to show that this should generally be the case. Conditions (1) and (2) are reasonable requirements [indeed, a violation of (2) might be worse than a singularity]. Condition (3) can be checked observationally and it holds. Condition (4) holds everywhere we have tested it in the Universe, but it will very likely break down when we get within 10^{-43} seconds of the singularity. Condition (5), represented by the inequality (*) concerning the density and pressure of matter, is the most interesting. It is usually called the *strong energy condition*. In the period from 1967 to 1978, before particle physicists hit upon the ideas of asymptotic freedom and gauge theories, which enabled them to extend their theories to very high energies, the inequality (*) was regarded as well founded.

As a result, the singularity theorem was widely cited as a mathematical proof that our Universe possessed a 'beginning'. Note that the theorem does not have anything to say about 'why' there was a beginning, or allow any investigation of the concept of 'before' the beginning: everything, all space, time, and matter, comes into being out of nothing at this singularity. It is the edge of the Universe. Strictly speaking, the singularity is not part of the Universe.

However, by the start of the 1980s particle physicists had discovered that their new theories abounded with matter fields (so called 'scalar' fields) which explicitly violated the energy condition (*). Alan Guth[6] showed that these fields can control the expansion of the Universe in its very early stages. While this control lasts they accelerate the expansion to create a phenomenon that Guth named 'inflation'. If the scalar field then decays into radiation and other particles which obey the inequality (*), inflation will cease and the Universe will resume its normal state of decelerating expansion, as shown in Figure 40.3.

Inflation was enthusiastically embraced by cosmologists because it offered a simple solution to a number of long-standing cosmological puzzles: why the Universe expands so close to the critical divide separating open from closed universes, why the Universe does not contain large numbers of magnetic monopoles, why the Universe expands so isotropically, and what determines the form and amplitude of the small density irregularities that grew by the process of gravitational instability into the galaxies and clusters we see today. Inflation offered a simultaneous solution to all these problems. As the structures of inflationary universes were explored in greater detail by Andreí Linde and others[7] it became apparent that it required a considerable enlargement of our thinking about the nature of space and time and the complexity of the Universe. It also brought bad news for provers of singularity theorems.

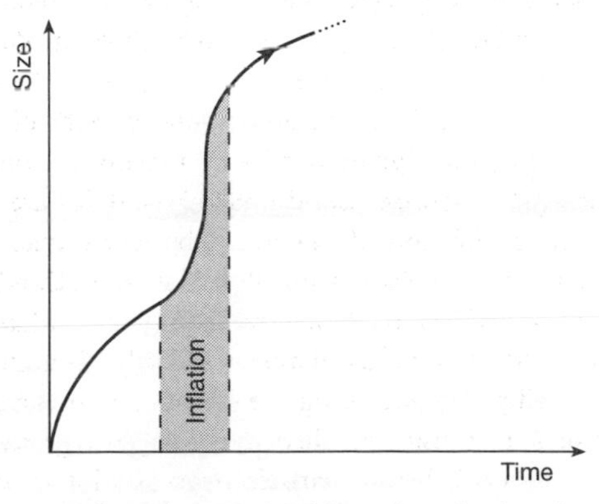

Fig. 40.3. The change in the early expansion history brought about by a finite period of inflation in which the expansion is accelerated.

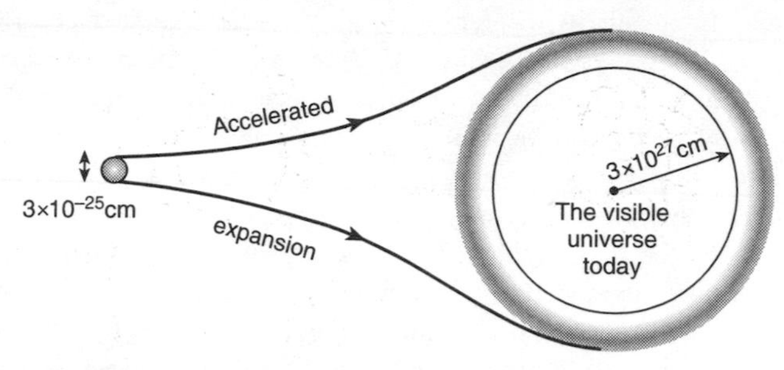

Fig. 40.4. The effect of inflation on a patch of the early universe that is small enough to be traversed by light signals at a representative early time, like 10^{-35} seconds. This patch expands to become larger than our visible universe today.

A necessary condition for inflation to occur turns out to be the requirement that the energy condition (*) of the Hawking–Penrose theorem be *violated*.[8] Thus, the singularity theorem does not apply to inflationary universes. We cannot draw any conclusions about the beginning of an inflationary universe: some inflationary universes begin with a traditional Big Bang singularity of infinite density, others need have no beginning at all.

Inflation implies that the entire visible universe is the expanded image of a region that was small enough to allow light signals to traverse it at very early times in the Universe's history (for example, at approximately 10^{-35} seconds in the most favoured model). However, our visible part of the Universe is just the expanded image of one causally connected patch approximately 10^{-25} cm in diameter (see Figure 40.4). Beyond the boundary of that little patch lie many (perhaps infinitely many) other such causally connected patches, which will all undergo varying amounts of inflation to produce extended regions of our Universe that lie beyond our visible horizon today. This leads us to expect that our Universe possesses a highly complex spatial structure and the conditions that we can see within our visible horizon, about fifteen billion light years away, are unlikely to be typical of those far beyond it.[7,9] This complicated picture is usually termed 'chaotic inflation'[7] (see Figure 40.5).

It has always been appreciated that the Universe might have a different structure beyond our visible horizon. However, prior to the investigation of inflationary universe models this was always regarded as an overly positivistic possibility, often suggested by pessimistic philosophers, but which had no positive evidence in its favour. The situation has changed: the chaotic inflationary universe model gives a real reason to expect that the Universe beyond our horizon differs in structure to the part that we can see.

Linde then showed that the situation is probably even more complicated. If a region inflates then it necessarily creates within itself, on minute length scales,

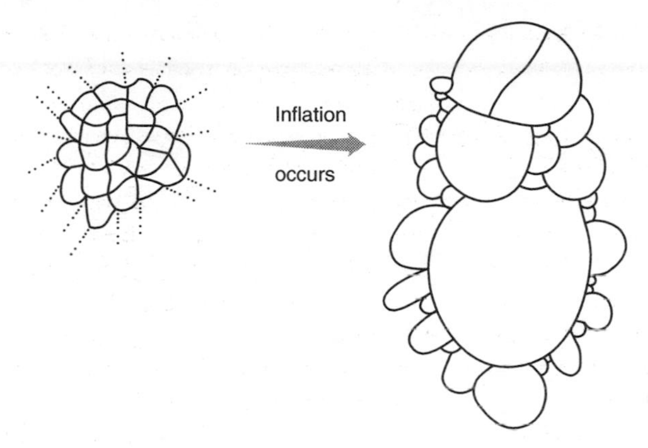

Fig. 40.5. Chaotic inflation. Different small patches of the sort pictured in Fig. 40.4 can undergo different amounts of inflation. We must inhabit a patch that becomes large enough to produce stars and living observers. Beyond our horizon the Universe should display a very different structure. Light has not had time to reach us from these regions during the expansion history of the Universe.

the conditions for further inflation to occur from many subregions within. This process can continue into the infinite future with inflated regions producing further subregions that inflate, which in turn produce further subregions that inflate, and so on . . . *ad infinitum*. The process has no end. It has been called the 'eternal' inflationary universe[7] (Figure 40.6). As yet it is not known whether it need have a beginning. But it looks as if a beginning is not inevitable for every region.

This enlarged conception of the inflationary model did not set out to produce an elaborate and sensational picture of the Universe. The self-reproducing character of the eternal inflationary universe seems to be an inevitable by-product of the sensitivity of the evolution of a universe to small quantum fluctuations in density which must exist from place to place.

Remarkably, if our visible universe has experienced a period of inflation during the very early stages of its history (the brief period from $t \sim 10^{-35}$ seconds to $t \sim 10^{-33}$ seconds suffices) then detailed traces should have been laid down in the structure of the microwave background on small angular scales. So far the COBE satellite has made observations on large angular scales which are in good agreement with the predictions of the simplest inflationary models. But future space missions by the MAP and the Planck satellites will map the microwave background fluctuations on smaller angular scales in minute detail and test the detailed predictions that inflationary models make. The past inflation of a patch that contains our visible universe is therefore a falsifiable scientific theory.

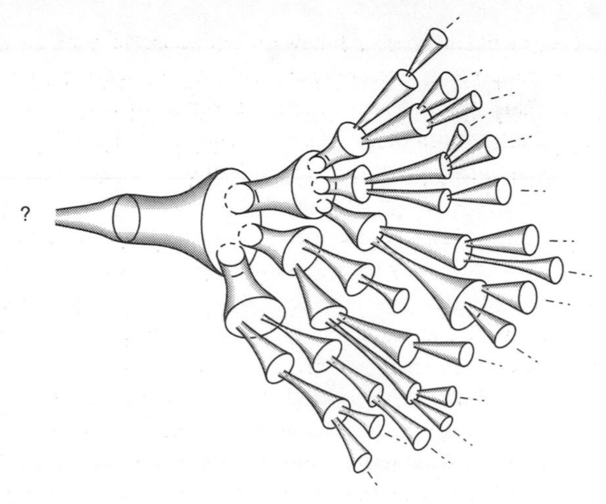

Fig. 40.6. Eternal inflation. The inflation of a small patch of the sort shown in Fig. 40.5 creates conditions within the inflated patch that initiate further inflation of subregions within it. This process appears to be self-perpetuating. It need have no end and may have had no beginning.

We should appreciate that inflation has both advantages and disadvantages as an explanation of the large-scale structure of the visible universe. It offers an explanation of this structure by means that do not depend significantly on the state of the Universe prior to the occurrence of inflation. Thus, all memory of conditions prior to inflation is lost. This has the advantage of allowing us to explain the structure observed in the Universe today without any need for knowledge of cosmological initial conditions (which might prove unknowable). But it has the disadvantage of preventing us learning anything about those initial conditions from astronomical observations today. Inflation wipes the sheet of creation clean and overlays it with a new structure generated during inflation.

If we look back further into the era before inflation may have occurred we soon run into another fundamental barrier to prediction. At a characteristic 'Planck' time, $t \sim 10^{-43}$ seconds, the non-quantum description of gravitation provided by Einstein's theory of general relativity must break down. A new, as yet unfound, theory of quantum gravity is needed to determine whether quantum cosmologies have a beginning. So far there are several complementary approaches to this problem. Some investigators have attempted to find a quantum wave function for the whole universe and to argue that a natural initial state exists for the Universe if we hypothesise that time becomes a fourth dimension of space when the Universe is small and dense.[10] Thus, time is viewed as a secondary concept that gradually becomes more persistent as the

Universe expands and cools. As we follow the Universe backwards towards the Planck era, the concept of time gradually melts away leaving us without the traditional Big Bang singularity of infinite density at the beginning of time. Instead, these quantum cosmologies raise new and unusual questions. One can entertain the possibility that the theory will provide a description of the appearance of the Universe (without physical cause) out of nothing by quantum mechanical tunnelling. The interpretation and reality of this phenomenon is still far from clear (and may never be) but theirs is one simple observation about it that is worth making. To anyone with a little knowledge of physics the idea that there could be a physical description of the inception of the Universe out of nothing seems impossible. After all, there exist absolutely conserved quantities of Nature—mass–energy, electric charge, and angular momentum. Surely the creation of a Universe out of nothing would violate these conservation laws governing these quantities and so such a process would always lie outside of a physical theory that preserves these conservation laws. However, when we inquire further into the nature of the mass–energy, charge, and angular momentum of our Universe we are in for a surprise. The mass–energy and charge must be precisely zero and observations limit any possible rotation (and hence angular momentum) of the Universe to be less than between 10^8 and 10^4 times less than the rate at which it is expanding. Thus, all these conserved quantities that make precise the idea that you cannot get something for nothing may well be zero in our Universe. Their conservation therefore presents no barrier to making a Universe out of nothing and describing it using physical laws that contain conservation laws. This feature of the Universe is surprising. Its full significance is as yet unknown.

We must end on a pessimistic note. There are really two 'universes'. On the one hand there is the entire (possibly infinite) *Universe*. On the other hand there is a finite region, roughly spherical in shape, and about fifteen billion light years in diameter. This is what we define to be our *visible universe*. It is finite in size. Its boundary is marked by the region of the Universe from which light signals have had time to travel during the past history of the Universe. There may be lots of Universe beyond the horizon of our visible universe—the inflationary universe theory leads us to expect that things will be very different beyond that horizon, but we will never know. Figure 40.7 shows a diagram of space and time on which our visible universe constitutes the cone of light rays travelling towards us from the past. All that observational astronomy can do is build up a picture of the structure of the boundary and the interior of that 'light cone'.

Our observations of the Universe are seriously and fundamentally limited by the finiteness of the speed of light. We can only gather evidence about the past history of our visible part of the Universe. Even with perfect instruments our

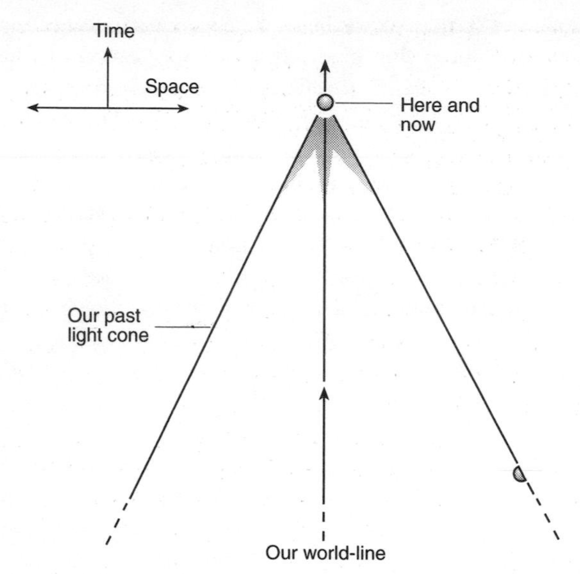

Fig. 40.7. Our past light cone. This delineates the region of the entire Universe from which we can receive light signals or massive particles. The region outside this cone is not accessible to astronomical investigation and we know nothing of its structure or existence.

picture of the Universe must be incomplete. We can investigate whether our visible part of the Universe had an 'origin'; but of the entire Universe we will always be ignorant. We cannot know whether it had an origin, whether it is finite or infinite. The information we need to make statements about these fundamental properties of our Universe is not accessible to us. The finiteness of the speed of light is one of the things that makes life possible in the Universe. It may be one of the things that makes the Universe possible. Ironically, it is also one of the things that prevents living observers knowing the Universe's ultimate secrets.

References

1. J.D. Barrow, *The origin of the universe*, Orion, London (1994); J. Silk, *A short history of the universe*, Scientific American Library, W.H. Freeman (1994).
2. J.D. Barrow and J. Silk, *The left hand of creation: the origin and evolution of the expanding universe*, (2nd edn), Oxford University Press, New York (1993).
3. R. Penrose, *Phys. Rev. Lett.* **14**, 57 (1965).
4. S.W. Hawking and R. Penrose, *Proc. R. Soc. London A* **314**, 529 (1970); S.W. Hawking and G.F.R. Ellis, *The large scale structure of space–time*, Cambridge University Press, Cambridge (1973).

[5] S.W. Hawking and R. Penrose, *The nature of space and time*, Princeton University Press, Princeton (1996).
[6] A. Guth, *Phys. Rev. D* **23**, 347 (1981).
[7] A. Linde, *Inflation and quantum cosmology*, Academic Press, New York (1990).
[8] J.D. Barrow, *Q. J. R. Astron. Soc.* **29**, 101 (1988).
[9] J.D. Barrow and F.J. Tipler, *The anthropic cosmological principle*, (2nd edn), Oxford University Press, Oxford and New York (1996).
[10] J. Hartle and S.W. Hawking, *Phys. Rev D* **28**, 2960 (1983); S.W. Hawking, *A brief history of time*, Bantam, New York (1988).

CHAPTER 41

Battle of the giants

The universe is real but you can't see it. You have to imagine it.

Alexander Calder

General relativity and quantum theory have always held a special fascination for physicists. They govern empires that appear superficially disjoint and rule their separate dominions with a precision unmatched by any other products of the human mind. The accuracy of Einstein's general theory of relativity, for example, is demonstrated by the spectacular observations of a pulsar engaged in a gravitational pas de deux with a dead star. Einstein's expectations are born out by observations—accurate to one part in 10^{14}. Almost as impressive is the accuracy of the quantum theory: agreeing with experiment at one part in 10^{11}.

The quantum world deviates strongly from that of Newton when things are very small. By contrast, general relativity only changes Newton's predictions when gravitational fields are strong and masses are very large. These conditions rarely overlap except in the cosmological problem of the Universe's first expansive moments.

Over the past thirty years, Stephen Hawking and Roger Penrose have done more than anyone to further our understanding of the nature of gravitation and cosmology. Both have developed new approaches to these problems that differ from the mainstream work by particle physicists (and from each other). *The nature of space and time* is the result of their attempt to stage a structured dialogue about these problems, to isolate points of disagreement, and stimulate further investigation of these problems. Alternate lectures are presented by the two protagonists, culminating in a final debate where they summarise their points of agreement and disagreement. The level of argument is highly technical, but you can skip the equations and still get a feel for what is going on.

Generally, great debates in science don't work. Science is not a democratic activity in which the idea that gains the most popular votes wins. Politicians

A review of *The nature of space and time*, by Stephen Hawking and Roger Penrose, Princeton, 1996

need not apply. Nonetheless, this volume shows that this adversarial style can be extremely valuable—at least at a textual level. Both authors are well acquainted with each others' ideas and write with great clarity. They agree on much and have to struggle a bit to play up points of disagreement over the interpretation of quantum mechanics, irreversibility, violation of CPT in gravitational collapse, the equivalence of black and white holes.

The opening two lectures introduce the minimum collection of ideas from differential topology needed to understand what a singularity is (an edge of space–time found, for example in a black hole, where all laws of physics break down), and the conditions under which it would be inevitable in our past. Then they move on to quantum effects and gravity with Hawking discussing black hole thermodynamics and introducing Euclidean methods. Penrose lets the cat out of the bag (and into a box) introducing the problems of interpreting quantum measurement, even proposing a simple formula for the time duration of wave function collapse.

The final pair of lectures is about quantum cosmology. Hawking argues for the inevitability of the Hartle–Hawking 'no-boundary' condition as a way of describing the initial state of the Universe which uses quantum mechanics to explain how time originates at the big bang. Penrose, on the other hand, argues for some measure of gravitational entropy. In this picture the second law of thermodynamics implies that a low value for gravitational entropy at the initial state would be natural. So the Universe would be almost isotropic and homogeneous initially, but chaotically irregular at any final singularity.

If you cast a critical cosmological eye over the proceedings, then several things are evident. Neither author is very impressed by superstring theory (to the exasperation of at least one questioner at the end of the lecture), both have wonderful geometrical intuitions and their taste in theories is strongly influenced by that penchant. Neither believes that inflation—the fashionable idea that the Universe underwent a phase of accelerated expansion in the first moment of its existence—is the whole answer to the problem of why the Universe is so isotropic and homogeneous today. And neither adequately considers the possible impact of cosmological inflation upon their viewpoint.

Hawking argues that superstring theories make no observable predictions that are not those of general relativity, to which superstrings reduce when gravity is weak. By contrast, he also claims that the no-boundary condition of quantum cosmology makes two successful predictions: the amplitude and spectrum of fluctuations in the microwave background. This claim is surely a piece of gamesmanship. These two predictions come from inflation, not from the no-boundary condition. The no-boundary condition allows inflation to occur but only by adding extra matter called a scalar field. This leads to the observed spectral slope of the fluctuations in microwave background radiations, but I could just as well have inserted a different scalar field into the early Universe which would give fluctuations in conflict with the observations, even though the no-boundary condition still holds. The 'correct' fluctuations come in all cases from an arbitrary choice of the scalar field, rather than from any prediction of the no-boundary condition itself.

Another important aspect of the picture of an inflationary Universe that both authors ignore is that observation requires only the 'beginning' of the Universe be uniform over a tiny region. Inflation can enlarge that uniform region so that now it is almost uniform over a region larger than our entire visible Universe. Beyond our horizon, however, the Universe could be quite different. The global structure of the Universe today may be extremely irregular: parts may be collapsing, rotating, or possess huge variations in density. This possibility arises naturally from general initial conditions.

The possibility of such initial conditions cuts through many of the assumptions made by both protagonists in this debate. They both maintain that current observations require a high level of uniformity in the initial stages of the Universe, using this to justify their own strong theories about cosmological initial conditions. But the observations do not require this and the initial state may have been globally highly irregular, contrary to Penrose's claim that the initial Weyl curvature was very small or Hawking's claim that the Universe began in a ground state defined by the no-boundary condition.

Considering this possibility is vital because, if allowed, it changes the entire nature of the Universe. It removes the evidence for any initial state of low 'gravitational entropy' or the need to distinguish fundamentally between initial and final singularities. Indeed, there need be neither a global initial singularity nor any quantum tunnelling of the Universe out of nothing. It shows how cosmology is unlike any other physical problem: the causal horizon structure of the Universe forbids us access to the information that we require to test a theory of cosmological initial conditions.

The debate between Hawking and Penrose is a live one between brilliant scientists that covers far more ground than their seven cameo lectures can encompass. This elegant little volume provides a clear account of two approaches to some of the greatest unsolved problems of gravitation and cosmology. It is recommended to critical readers, who should not forget that there are other more widely supported views about these cosmological problems. Which, if any, view is true? At present only God knows—or maybe not.

CHAPTER 42

They came from inner space

Your manuscript is both good and original; but the part that is good is not original, and the part that is original is not good.

Samuel Johnson

Disease is ubiquitous. Even the optimists who seek the creation and evolution of artificial life have found their silicon beings attacked by viruses that multiply and spread throughout their software. Computer buggery is in the business news but carbon-based lifeforms have seen it all before. Not a week passes without a new scenario for the 'Death of the Dinosaurs' (now showing at a cinema near you). From 'Silent Spring' to 'Nuclear Winter' the seasons may change but the interaction between life and its wider environment is a never-ending story of unsuspected connections between the inner space of biological functioning and an outer space teeming with unwanted visitors. Innocent victims like the dinosaurs have died a thousand deaths from the impact of meteorites, climatic change, comets, and kindred disasters.

Until now the consequences of this patchwork of unrelated ideas for the long-term health of our planet have looked like the pieces of a dozen muddled jigsaw puzzles sitting in the kids' bedroom waiting to be sorted out. But when the sorting started a number of vivid and alarming pictures emerged. In recent months scientists in many countries have begun to piece together a startling scenario that threatens to revolutionize our ideas about how, when, and where diseases originate and why our counter-measures to prevent their spread are so often unsuccessful.

We have always regarded the origin of disease as a skein of many strands, knotted and tangled, without a common thread, and lacking a common source. But we have repeatedly seen how the other physical sciences, most notably physics, have made dramatic progress by searching out unified descriptions of the phenomena they investigate. No more do physicists see the electro-

By *'Rolf Paoli'*

magnetic, radioactive, nuclear, and gravitational forces of Nature as independent; rather, they are understood as different manifestations of a single grand unified force. When this unity is realized many of the idiosyncracies of the individual forces dissolve in our new realization of the role they play in the harmony of the unification process. It is this unification philosophy, together with the increasing link between the physics of elementary particles and the astronomical universe, that has fuelled fast-moving developments in medical research. A number of young physicists, epidemiologists, and biochemists have begun to see exciting new potential at the new and relatively unexplored interface between physics and epidemiology.

For many years now astronomers have known that there is a 'dark side' to the Universe. Gravity reveals the presence of between twenty and a hundred times more matter in the Universe in dark form than can be detected by its emitted light from all the stars and galaxies. The presence of so much 'dark matter' is not necessarily surprising; the formation of stars and galaxies may simply be a rather tricky business which fails more often than it succeeds. But what is astonishing is the discovery that ordinary forms of matter that you (I assume) and I are made of are something of a rarity in the Universe. In order to be consistent with other astronomical observations this dominant dark matter must be in the form of a more or less uniform cosmic sea of weakly interacting massive particles ('WIMPs') rather akin to the ghostly neutrinos that we are familiar with (millions are passing through your head as you read this) but a billion times heavier. Such particles are predicted by new supersymmetric theories of matter which the mega-billion dollar Superconducting Super Collider (SSC) is being designed to test. Today they are distributed within our Milky Way galaxy with about one to every cubic centimetre; each WIMP has a similar mass to that of small atoms and moves with a speed close to 250 kilometres per second, or 600 000 miles per hour. These particles impinge upon the Earth with a significant flux and produce a steadily accumulating spectrum of genetic damage and enhanced carcinogenic mutation. Many of these forms are novel because of the particular type of interaction the WIMP undergoes when it recoils from the atomic nucleus of a biological atom in a DNA chain. One possibility is particularly insidious. It is very likely that these particles interact with atoms via their intrinsic spin and so might be able to alter the intrinsic handedness of biological molecules directly. The devastating consequences of such seemingly innocuous changes can be appreciated by recalling that thalidomide resulted from the presence of a wrong-handed version of a harmless molecule in a sedative drug.

The magnitude of these effects are being studied by particle physicists in underground experiments which shield out the confusing (but quite different) effects of cosmic rays and local terrestrial radioactivity. Large numbers of fruit flies have been studied in tanks surrounded by batteries of particle detectors able to monitor the flux of incoming WIMPs so that the effects upon the mutation rate of the fruit flies can be quantified over many generations of their short life-cycle. With knowledge of this basic effect reasonably complete, medical physicists have focused upon the cumulative effect of the WIMPs that accumulate within the Earth as they are trapped by the pull of the Earth's gravity. Since the density of material is greater as one moves below the surface of the Earth the probability of interaction with other matter increases dramatically and the enormous energies carried by the fast-moving WIMPs is continuously deposited within the Earth. The result is a significant global warming which has been going on throughout the history of the planet. The warming is not continuous because of the 246 million year orbit of our solar system around our galaxy. Only in recent years has this become obvious because during some epochs a local deficit in the WIMP flux arises or the solar system passes through a dusty spiral arm of the Milky Way. Moreover, the red-herring of fossil fuel consumption has distracted the vast majority of climatologists from the true source of the problem.

The build-up of interactions within the Earth has left our species open to attack from within as well as without. Direct mutations caused by incoming WIMPs may turn out to be less important than the more debilitating effects of high-energy particles emerging from the centre of the Earth. In the past, when their rate was modest, they appear to have played a key role in the origin of life by stimulating biochemical diversity and mutation in the stable, moist, and temperate environment that is supplied by the surface layers of the Earth's interior. But today, the evidence of global warming reveals that the subsurface mutation rate is reaching levels that present an unprecedented danger to long-established surface life-forms like ourselves. In a bootstrap process reminiscent of a belligerent form of the GAIA hypothesis the subterranean heating by WIMP encounters with ordinary atoms triggers greatly increased vulcanism and plate tectonic activity. The fissures that result from plate bending and earthquakes allow huge numbers of complex self-reproducing molecules to come to the surface on land and sea. Their terrestrial effects have already been seen. Increased incidence of novel genetically complicated disease, like AIDS, to which we have no naturally evolved immunity is expected close to earthquake fault lines. This is clearly seen in northern California and in the Rift Valley regions of the African Continent where such diseases are rife.

Marine effects are less evident at present but ultimately they will prove no less devastating. Two-thirds of the Earth's surface is covered by oceans. The vast majority of invasive viral forms will enter the ecosystem and the food-chain through marine life-forms. Studies have shown that small organisms, for instance krill, which are eaten in astronomical abundance by large sea creatures like whales will gradually produce lethal variations and disease that will pass into human societies via fish. Ironically, in many Western countries the consumption of fish is growing dramatically at the expense of meat in response to worries about the effects of fats, animal rights lobbyists, and mad cow disease. The policies of such groups, together with the paternal attitudes towards large sea mammals like whales, may turn out to require a radical rethink. Large commercial interests are also at stake. Shellfish such as oysters process vast quantities of sea water, and pearls are seen as the most likely repository of huge numbers of potentially carcinogenic molecules. The danger is the more acute since these items are invariably worn next to the skin in large numbers.

The message from these new discoveries is a somber one that calls for concerted international action. Former British Prime Minister Thatcher and members of her cabinet were briefed in detail by a British scientist involved in the study of WIMPs over a year ago but no public statement has since been made by the British Ministry of Health. Nor has any statement been forthcoming from the Surgeon General in the United States. Action is needed. But just what could that action be? The most insidious thing about this threat to the planet is its unremitting universality. We seem to be taking on the whole Universe: WIMPs cannot be stopped and they are not going to go away. Worse still, their adverse effects snowball. As the global warming continues so the reproduction rate of carcinogenic mutations increases while the atmospheric ozone crisis leaves us open to increased carcinogenic effects from the direct bombardment by WIMPs and other cosmic particles.

All our efforts at stemming the AIDS epidemic have focused upon stopping its spread under the assumption that it had a single source. But now we recognise that it is constantly being rejuvenated in subtly different strains over the whole face of the planet. The World Health Organisation is soon to launch a global forum for scientists from many disciplines to pool their ideas as to the best course of action to take. All we can say with confidence at present is that one should avoid living in places that lie within about two hundred miles of a major earthquake fault and avoid sea bathing and contact with any form of jewellery like pearl, coral, or eel skin that has a marine origin. As the true scale of this crisis becomes the focus of public concern one can predict significant demographic changes. In the US and Japan the inhabitants of some of the most valuable areas of real estate now appear blighted by an emergent epidemic beneath their feet. What is to be done? It took Superman to stop Lex Luther shifting the peaks of the property value contours from California's coastal San Andreas fault line to Arizona and Nevada. Who will stop the WIMPs?

Biographical sketch: Rolf Paoli

Professor Paoli was born in Padua, Italy, in 1950 and is a graduate *summa cum laude* of the universities of Rome and Cambridge where he took Bachelors, Masters, and Doctoral degrees in physics and biochemistry. After a distinguished early career working in research groups at Brookhaven and the Lawrence Berkeley Laboratory studying the effects of natural and man-made radioactivity upon humans and animals his research interests changed direction in 1987. Since then he has worked in biophysics and on the role of the terrestrial environment in epidemiological studies. In 1989 he was appointed Accademia Lincei Research Professor at the International Centre for Biophysics in Trieste. He has published over 100 research papers and is a frequent commentator on environmental health matters in the European media. He is an editorial consultant to *Nature* and the founding editor of the *International Journal of Exoepidemiology*. Since 1990 he has been Chairman of the AIDS Audit Group of the World Health Organisation. Professor Paoli lives in Venice with his wife Polli Faro, a well-known literary scholar whose most recent work on the meaning of the opening words of T.S. Eliot's poetic work 'The Waste Land' is to be published shortly by Cambridge University Press in New York. They own a small black cat. Neither Professor Paoli nor his wife drink Dewar's Scotch whisky. Their cat does.

Index

"Brennan," his mother said, "I'd like you to meet Bill Halverson."

Brennan nodded. "Hi."

"Hi."

And that was it, or he thought it was. Brennan went into his room and changed quickly and threw on a T-shirt and made his way back outside to jog down to Stoney's place to ride in his old pickup out to mow lawns.

Later he would think back on this time; later when he had begun to try to find his spirit and see the dance of the sun, later when his life was torn to pieces and he was trying to make it whole again, he would look back on this moment, this exact moment when it started, and wonder how it could be.

How anything so big could come from something so small and simple.

"Hi."

"Hi."

And it changed his whole life.

GARY PAULSEN is the distinguished author of many critically acclaimed books for young people, including three Newbery Honor Books, *The Winter Room*, *Hatchet*, and *Dogsong*. He lives in northern Minnesota with his wife and son.

CANYONS

GARY
PAULSEN

Published by
Dell Publishing
a division of
Bantam Doubleday Dell Publishing Group, Inc.
666 Fifth Avenue
New York, New York 10103

To Angenette
and James Wright—with deep gratitude,
true love, and gentle respect
for the joy
of their life

ISBN: 0-440-21023-2

RL: 5.2

Reprinted by arrangement with Delacorte Press

Printed in the United States of America

September 1991

10 9 8 7

RAD

CANYONS

1
Quickening

Soon he would be a man.

Not after months, or years, as it had been, but in a day. In a day Coyote Runs would be a man and take the new name which only he would know because finally after fourteen summers they were taking him on a raid.

He had difficulty believing it. For this summer and two summers past he had gone time and time again to the old place, the medicine place, the ancient place in another canyon only he knew about, and prayed for manhood—all for nothing. For the whole of this summer and two summers past he had been ignored, thought of as still a boy.

And now it was upon him.

In the morning he had risen early and walked

away from the camp before the fires yet smoked and gone to the stream to rinse his mouth and take a cool drink and nothing seemed different.

He had gone to the ponies and looked at them and thought how it would be someday to own a pony, two of them, many of them, when he was a man and could go on raids to take horses and a Mexican saddle with silver on it like Magpie had done when he became a man and who could now walk with his neck swollen.

All someday. That's how he thought then, in the morning, it would all come someday. He would be an Apache warrior and ride down past the bluebellies' fort across the dirty little river into Mexico and prove that he was a man.

Someday.

After he had watched the pony herd for a time he walked back to the huts in the early morning sun and saw that his mother was making a fire to heat a pot of stew made from a fat steer they had taken from Carnigan's ranch. The rancher had many such steers and did not mind when the village took one. He liked to watch his mother make fire. She was round and her face shone in the sun and her hands were so sure when she piled the sticks and struck the white man's lucifer stick to make the fire that he thought of her as not just his mother but the mother of the fire. As she was the mother of the stew and the wood and the sand and the hut and Coyote Runs. The mother of all things.

Then Magpie had come out of his hut where he slept with his family though he was a man because he did not have his own wife yet. He saw Coyote Runs sitting near his mother and came to him and squatted next to him in the dirt.

"It is a fine morning," he said, which caused Coyote Runs to look at him because his voice was light and teasing. "A fine morning to be a man. . . ."

Coyote Runs said nothing but had a sour taste in his mouth. It was not like Magpie to make fun of him or to be proud so that it showed in a teasing voice.

"I have heard stories," Magpie said, and now Coyote Runs could tell that he was teasing openly. But he was smiling and his eyes were not mean.

"What kind of stories?"

"I have heard stories of a new raid to where the silver saddles are. A raid which will leave tomorrow. A raid which will have all the warriors on it."

Coyote Runs felt it now, the small excitement that came from surprises.

"All the warriors and men there are. Tell me, warrior, do you have a pony?"

"I'm going?" Coyote Runs tried to keep his voice even.

Magpie smiled wider and nodded. "It is thought that you could come to hold the horses and see how it is to be on a raid. Sancta said it, said you could come by name."

Coyote Runs thrilled inside but tried to remain cool, not show it, as was proper. For a man. As a man should act. Sancta was a scarred old man who could not be touched by arrows or bullets who had led all the raids since Coyote Runs knew there were raids; Sancta decided who would go and who would stay. And he had decided. "I do not have a proper pony for war but it is perhaps true that I have a friend with a pony." He looked pointedly at Magpie. "Do I have such a friend?"

Magpie nodded. "I will loan you that small brown pony with the white eye. Until you can get your own."

"Until I can get my own."

Magpie stood. "I think I will go down to the stream now and clean myself."

Coyote Runs nodded but did not stand. Inside he was ready to explode but he remained cool. "Yes. The water was good this morning."

And he thought I am to be a man. I am going on a raid and I am to be a man.

There was much to do. He must ready himself and his bow and check his arrows; he must make certain that everything was the best that it could be.

I am to be a man.

There was much to do.

2

Brennan Cole lived in El Paso, Texas, and each afternoon after school he ran. He did not run from anything and did not run to anything, did not run for track nor did he run to stay in shape and lose weight.

He ran to be with himself.

He was tall and thin and healthy with brown hair that grew thick and had to be cut often and because he ran in shorts with no shirt and wearing a headband instead of a cap and because El Paso sits at the base of Mount Franklin and burns in the hot sun for most of the year he was the color of rich, burnt leather.

He did not know his father. He lived alone with his mother and when he was home—which was less and less as he approached fifteen and his mother spent more and more time working to live, working to be, working to feed and clothe her only son—the two of them existed in a kind of quiet tolerance.

She did not dislike him so much as resent the burden she thought he was; and he did not dislike her so much as want to relieve her of the burden.

He was still too young for fast-food jobs but he mowed lawns for a lawn care service, working for an old man named Stoney Romero, who paid him in cash and did not ask questions nor give answers except to aim Brennan at yards with a mower. Brennan did not make much, but he fed himself and bought clothes for school and special shoes.

For his running, he thought now, turning on Yandell Street headed for the apartment over the house where he and his mother lived. I need money only for shoes for running.

It is all I need to do—to run is all there is. When I am running it is all, everything. Nothing matters. Not the father that I do not know or the mother that does not know me or the school that I hate or money or not money—all of that disappears when I run.

When I float. When I run and float. God, he thought, his legs felt like they belonged to somebody else, somebody who never became tired, and when he looked down at them pumping, driving, moving him forward they marveled him. From all the running, from the daily running in the streets and up the hills they had become so strong he didn't know them.

There was traffic but he keyed his steps and ran across the street at an angle, picking up speed as he passed in front of an army pickup from Fort Bliss. I can run for all of time, run forever.

It was May and very hot, close to ninety in the shade. But dry heat—always dry heat in the desert—and it didn't take him down. Nothing took him down.

Except being home.

There was that, he thought, turning up the block and seeing their apartment and his mother's old Volkswagen, knowing that he would not be able to be with himself any longer. Only home took him down.

She had company.

She belonged to one of those parents-without-husbands groups or whatever they were and from time to time she would bring different people around and he would have to meet them. He tried to be nice to them, but they always looked at him with open pity and he didn't feel that he should be pitied simply because he didn't know his father, had never met him, didn't know who he was, or any of the other ways they put it. The truth was he had never really had a father and so it didn't matter.

But she went to these meetings and she told everything about herself at the meetings, which included him, and everything about him, and so these people he didn't know knew all about him and would come to the house with her sometimes and sit and pity him.

They all wanted to "share," and "care," and "get in touch with their feelings," and on and on. The first few times they had come and met with his mother in their kitchen he had thought it might be all right and that his mother might get out of herself. But after a time, and many, many meetings, it just seemed that she was spending all her time messing with herself and not trying to really fix anything.

And I have to go in, he thought. I have to work for Stoney and that means I have to change clothes and get jeans on to go to work and that means I have to go inside. If it were not for that, he would run past the apartment and keep running for a half hour or so, until the company was gone.

Maybe if he just whipped in and changed, he could get away before he got nailed.

But it didn't work that way.

He ran up the back steps and into the kitchen, where his mother was sitting with a man.

He was pleasant enough. Tall, slightly heavy, with short hair.

"Brennan," his mother said, "I'd like you to meet Bill Halverson."

Brennan nodded. "Hi."

"Hi."

And that was it, or he thought it was. Brennan went into his room and changed quickly and threw on a T-shirt and made his way back outside to jog down to Stoney's place to ride in his old pickup out to mow lawns.

Later he would think back on this time; later when he had begun to try to find his spirit and see the dance of the sun, later when his life was torn to pieces and he was trying to make it whole again, he would look back on this moment, this exact moment when it started, and wonder how it could be.

How anything so big could come from something so small and simple.

"Hi."

"Hi."

And it changed his whole life.

3

Dust Spirits

Coyote Runs let the small brown pony pick its way down the side of a dry wash and thought that it was as his spirits, the spirits of the dust, had told him—everything was perfect.

The afternoon before they left on the ride south to Mexico he had gone to the ancient medicine place, the secret place, and had spoken to his spirits to ask for guidance and bravery to have a thick neck and be a man. He had waited for a long time and nothing came and he was beginning to worry that it was all wrong, that there would not be a sign, when it came:

Out below him to the east in a dry lake bed the wind swirled and picked up a column of dust and carried it heavenward, carried it to the spirits, carried his wishes high and away, as high and away as the hawk, as the dust, and he knew it would be all right. Would be perfect.

He painted the pony with one circle around one eye so that it could see well if they had to run at night and put tobacco on its hooves to make them fast, to show the spirits where he needed help.

Each arrow he placed tobacco on, using tobacco from a round metal tin that Magpie had found in an old shed at the Quaker school when the two of them had gone to the school to learn how it was to be white. They learned nothing except some symbols on a black stone written with a piece of white dirt; symbols that meant their names in white man's words that the Quaker lady taught them which did them no good because nobody else could read them. But Magpie found the round tin with the lid in the shed so it was not all for nothing.

He did not have a gun. Some of the men had guns and all the soldiers had guns but he did not have a gun and would have to use the bow and the arrows and when he had put tobacco on each arrow he did the same for the bow, each time asking for the dust spirits to make them fly, make them shoot in the proper manner.

All afternoon he prepared himself and when the men sat that night and talked of the raid outside one of the huts he stood to the rear and listened. He thought for a moment that Magpie had been teasing him because two of the men looked at him. But when they did not tell him to leave he knew that he was to go on the raid, that Magpie had been telling him the truth, and his heart was full of joy.

It was the first time he had heard them talk of a raid when he was not sneaking and he saw other boys who were still too young hiding round the back of the firelight in the dark and felt pride that he was at last allowed to be with the men. He listened carefully, quietly, with great courtesy and said nothing.

"We will take extra horses in case there is an injury and ride to that place on the river where the land cuts low so the bluebellies cannot see us cross when we go or when we come back," Sancta said. He was wrinkled and had scars from many battles. There was a line across his forehead which was said to have come from the long knife of a bluebelly who had tried to cut the top of his head off and let the light into his brains.

And the men grunted and nodded and smoked in silence, spitting on the fire from time to time.

"We will stay tight in together until we get to that place below the cut on the river where the Mexican ranchers keep that big herd of horses for selling to the bluebellies."

More grunts and nods.

"There we will leave Coyote Runs with the extra horses and hit the herd and take as many horses as we can get running."

Now there were open exclamations and Coyote Runs felt a thrill of pride that he had been named by name as going on the raid. He turned to see if any of the hiding boys had heard but he could not see their faces, just shapes in the brush and darkness.

It did not matter, he thought then and thought now as his pony followed the rest of the men and extra horses. All that mattered was that the dust spirits were with him and had heard his request and that he was on the raid.

It would take three days riding out and around to get to the river that the white men used as a boundary and part of another day to ride south to where the Mexicans kept the big herd of horses and then two days to ride back driving the horses ahead of them. Six days, Coyote Runs thought, riding easily.

Six days to test himself and prove he was a man.

Surely there would be many ways for him to be a man in a week.

He used his heels to goad the brown pony up and out of a gully Sancta had led them to; ahead Magpie caught the sun and his long black hair shone like a raven's wing in the light and Coyote Runs smiled at the beauty of it.

Six whole days, Coyote Runs repeated. Six days to be a man.

4

"You must come in closer on the flower beds or there is too much for me to use the string on."

Stoney Romero gave him instructions the way he did all his other talking—his voice like gravel rattling around in a garbage can. He smoked cigarettes which he rolled from tobacco in a small plastic bag he carried because "tailormades" cost too much and he spent most of his time coughing and hacking.

"If the mower gets too close I'll take out the beds," Brennan explained. "And then you'll get chewed out and I'll get chewed out."

"Nobody chews me out," Stoney told him, and it was true. Brennan had never seen anybody speak crossly to the old man. Perhaps because he was scarred across his cheeks. A part-time worker also named Romero had told Brennan that it was from a knife fight in prison where Romero killed a man, but the part-time worker also told him that Fig Newton cookies could cure baldness, so Brennan nodded. "I'll work in closer."

"As I said."

Stoney turned away and Brennan started the riding mower and began to mow. He liked the work, liked the way the mower worked in rows and cut even, fresh green lines with each round. Stoney had many lawns to mow for the people he called "the rich ones."

It wasn't even that they were all rich, although they did some lawns on houses that were huge old estates where there were statues around the pools and steel gates that had to be opened electrically. Many of the lawns belonged to army families who did not make as much money as Stoney. But they were all "the rich ones" to him, said with a sharpness to it that meant he did not respect them, and Brennan was glad for the work.

He needed the money. His mother had told him she couldn't afford to help him with his school clothes as much as she'd thought she would because she didn't get a raise she thought she would get. Brennan had tried to get other jobs but he was too young. They didn't care if he could do the work or not. He was too young.

And Stoney hadn't even asked about his age.

"Can you mow a lawn?" he'd asked, taking a deep drag on the small brown cigarette and coughing.

"Yes."

"You're hired. I will pay you three dollars an hour in cash at the end of each day and if you do not show up for work you are fired."

And he had been as good as his word. Usually work was almost impossible to find because of the closeness to Mexico—Juárez was right across the river—and the poverty there which sent thousands north each day to find work.

But Stoney had hired him and he'd worked now for almost all of June and July and Stoney had been as good as his word. Each day he paid Brennan in wrinkled bills and quarters and dimes and nickels, exactly the amount for the hours he had worked and each day Brennan went home and put the money in a jar in the cupboard.

It was adding up, the money.

Because I don't do anything, he thought, moving the mower in closer to the flower beds as instructed, squinting up at the hot sun and the Franklin Mountains that rose above the city—except work and sleep.

He had no really close friends, no girl—just himself. It was strange but he didn't seem to make connections with other kids. Last year for most of the year he'd been close to a boy named Carl and he guessed Carl had been his best friend. But Carl didn't run and Brennan did and after a time they just drifted apart and sort of stopped calling each other and that was that.

Some of the jocks wanted him to join the track team, as did the coach, Mr. Townsend. But they ran for the wrong reasons as far as Brennan was concerned. They ran to win, to be somebody, to get popular, to get girls, to get a letter—not just to run. Brennan ran for the joy of running, and to be with himself, and when he told the coach that, Mr. Townsend had looked at him as if he were crazy.

"You're good," the coach had said, "really good. You could help the school, make something of yourself, be popular. . . ."

But Brennan could not see why it was important to do those things and when the coach saw that he only wanted to run and when the jocks saw that he only wanted to run and the other kids saw that he only wanted to run they left him alone.

To run.

He worked in close around the flowers. Stoney was right, he could get closer. But it was dangerous and if he took out a flower bed, or even cut the edge off one they would probably lose the job. And he might lose his job as well.

So he worked slowly and carefully and the afternoon went slowly. They finished the lawn with the flower beds, then another one which wasn't too far away and then they loaded into Stoney's old pickup to drive to a third one, which would be the last for the afternoon.

On the way, out of nowhere, Stoney decided to talk to him.

"Kid."

Brennan looked at him. Generally, other than giving instructions, Stoney treated Brennan as if he weren't there.

Just sat smoking and scratching or working by himself and left Brennan alone to his own thoughts.

Brennan looked across the truck at him. "What?"

"Ever think how stupid this is?"

"What?"

"This whole thing with the lawns. Here we are in the middle of a desert where nothing grows but mesquite and cactus and some snakes and people spend all their money on water just to get grass to grow so they can spend even *more* money to get us to cut it for them so they can spend even *more* money on water to get it to grow again so we can cut it. . . . Isn't it stupid?"

"Well, I guess so. But a lawn is nice. A green one."

"What does your old man do?"

The question came fast, dropped out of nowhere, but Brennan had answered it so many times, the answer was always ready. "Nothing. He died when I was a baby."

The lie. It came so easy now and he had done it so many times that he almost believed it himself. He died when I was a baby. It was so sad. He died when I was a baby.

The truth was, he had run off with another woman when Brennan was three and Brennan could only just remember a faint image of how he had looked—like an old photograph blurred and faded with age. For a time he had hated his father for leaving, but even that was gone. Now there was nothing.

"Tough." Stoney coughed and spit out the window. "Having no father . . ."

"It's not so bad," Brennan said quickly, too quickly. "Mom and I get along."

"Well, hell, it might have been worse the other way. My old man lived and spent all his time sucking on a tequila bottle and beating the snot out of me. There were a few times I wished he would die, come to think of it. Maybe if your old man had lived he would have beat you."

Brennan didn't answer, sat looking out the window. The heat coming down from the mountains, the sky, up from the asphalt, was like a blast furnace. It made the metal on the truck door so hot that if he moved his arm to a new place it burned him. But he liked the heat just the same. There was something clean about it, fresh. Yeah, he thought, if he'd lived he probably would have beat me. But even that might have been better than what they had now. Now it was so . . . so nothing. He and his mother just spent all their time in the house getting along. Just a day and then another day and then another getting along.

And her friends, of course. There were her friends.

Like this guy Bill.

He wondered if that would turn into another relationship. His mother had gone through so many relationships, he had lost count. Dozens. Well, maybe not dozens. But many.

Some of them had moved in and he had played the game with them. Treated them like substitute fathers. But he no longer did that. He was too old for it and it just felt silly. Most of them had been good to him, some very good. One named Frank who showered him with presents and toys and candy and took him places and was almost sticky about it. But even with Frank he had trouble thinking of him as a father. It wasn't the same as having a real one. Or at least that's how he thought.

They unloaded the mowers and mowed the last lawn. It was large and they finished in the evening, just before dusk.

"You want a burger?" Stoney asked while he paid Brennan. It sounded like an invitation but Brennan knew that what it meant was that Stoney would take him by a McDonald's where he could buy his own burger and eat it on the way home in the truck. Brennan shook his head.

"No. I'd better get home. Mom is waiting." Which wasn't true but served as well as the truth and Stoney nodded and drove him home in silence.

Brennan saw that his mother's car was still parked in front of the house, along with another car he had seen earlier that day and when he walked in he saw Bill sitting at the kitchen table with his mother.

"Oh, Bren," his mother said, her eyes wide and smiling. "Come on in here. We've got great news. We're going camping with Bill. . . ."

Oh great, Brennan thought, another relationship.

5
Nightride

The village lay east of the white man's town of Alamogordo three, four days riding but they did not ride toward that place. Coyote Runs knew there was nothing in that place but dusty little houses and a place where men drank until they fell down and puked. He had been there once when he had been going to the school with the Quaker lady and was sickened by what he had seen.

The raiding party would not go near Alamogordo because they might be seen. Instead they rode straight south over the tops of the wild canyons that fed out into the great desert and the place with the white sand the Mexicans called Jornada del Meurto—the journey of death. It was said that a Spanish warrior in a time long ago before there were even horses except for those he brought rode through that place and lost many men and had to eat some horses but Coyote Runs did not believe all of that.

There were many stories of hard journeys told by old people, some of their own stories and some they'd heard from the Mexicans, but he did not find them all truthful. He had eaten horse himself many times and it was not so special a thing to do; the meat was good and the fat was yellow. If there was fat. But it was a silliness, a stupid thing to do—eating a horse. There were better things to eat and you would not lose your ride.

Coyote Runs shook his head. All this thinking was a waste. It was just that it was truly dark now and Sancta continued the ride, letting the horses pick the trail as they made their way south. There was nothing to see and Coyote Runs rode in silence, as did Magpie in front of him and all the others, and without anything to see or anybody to speak to or listen to his thoughts went a way of their own.

Like thinking of the place of white sands. He had been to it only once though he could see it from the medicine place and thought it looked like the snow that covered the mountains in the winter but when he touched it the sand was hot. White with the sun and hot. Like hot snow. Could there be such a thing, he wondered—such a thing as hot snow?

His pony stumbled and he grabbed its mane to

pull it upright. Some horses could see better in the dark than others and he hoped that Sancta's horse was one of the good ones or they could all ride off a cliff. There were many steep drops to the right and he knew that if he could see it would be possible to see across the desert to the other range of mountains, the jagged-tooth mountains that the white men called the Organ Mountains because it reminded them somehow of the music box the Quaker lady had in her home that required you to pump your feet to make the music come out the pipes. The music had been nice, but naming the mountains after the box made no sense at all to Coyote Runs. That mountains could look like the small box, mountains that reached to the sky with arms of stone, could make the white men think of the small squeaky box in the Quaker lady's home was a silliness.

He shook his head again. About as much a silliness as riding along in the darkness on his first raiding party thinking small thoughts when he should be thinking of becoming a man.

He must think of that as he rode. Becoming a man. He must ride straight and think of serious things, think of the raid and not being afraid if he was called upon to do battle or if they met the bluebellies.

He straightened and tried to see ahead in the darkness, tried to pay close attention to where he thought Magpie was moving ahead of him.

He must be serious, as serious as a man must be.

6

Brennan looked out the window of the van and worked very hard at not being angry. They were driving north of El Paso on the highway that went through the desert up to Alamogordo and he sat looking sideways out at the huge rock canyons that led up from the desert into the mountains. They were beautiful, colored layers of rock and natural formations that looked painted and yet he was having trouble seeing any of the beauty in them.

He'd somehow gotten cornered on this camping trip thing and he was still trying to figure out how it had happened.

Normally he was really good at avoiding becoming involved in his mother's business. When she got into aerobics he avoided it and when she did the health food he avoided it and once he'd even avoided it when she was going with a jock who thought he knew all about running and actually spent time running with Brennan, running alongside him when he ran, telling him how to hold his arms and move his legs and how to breathe. That time Brennan had outrun him—just hung a right and headed up toward the drive around Mount Franklin, loping along and letting the guy fall back until he was just a speck.

But sometimes you lose, he thought, looking at the scenery blur by—sometimes no matter what you do you lose.

And this was one of those times.

Two days, no, three days ago he'd been standing in the kitchen and his mother had been smiling and saying:

"We're going camping with Bill and his youth group."

And his mind had raced through a thousand excuses, some new, some old. She had started going to this new church and she'd met Bill at a church dinner and Bill was active with a youth group—she had explained it later to Brennan, her cheeks glowing. But he hadn't known that at the moment, at the second, the instant when he thought of all the excuses. I'm busy, I have to work, Stoney has a lot of new lawns—the cards flipped through in his mind and he opened his mouth to use one of them but there was such a hopeful look in his mother's eyes, such a childlike hopeful look that the words that came out had nothing to do with what he was thinking:

"Oh? That will be nice."

Dead, he thought looking out the window of the van. I was dead when I opened my mouth.

And once he'd said that he couldn't get out of it. Not really. And his mother *had* wanted it so badly, or seemed to.

And he'd thought, oh well, it won't be so bad. Overnight out in the desert with Bill and his mother and some kids from Bill's church youth group. How bad could it be?

He almost smiled now, thinking back on it, would have smiled except that it *had* turned out so awful. The kids—there were seven of them, all boys, all about eight years old—were monsters. They were all over the van like gremlins, wouldn't let themselves be buckled in, and with Bill and Brennan's mother in the front seat and Brennan in the back with the kids—he thought of them as the pack—the main load of work with the children dropped on Brennan.

It was like being in a nest of rats. They climbed on the seats, bit each other, fought, and wouldn't do anything Brennan told them to do. One of them, a boy named Ralph Beecher, just sat in the corner of the backseat kicking anybody who came within range.

By the time they were out of El Paso, Brennan knew he was in trouble and within twenty miles of the highway heading north he had his hands full. He tried holding them down, pushing them away, scowling, swearing at them—nothing worked and finally he turned away and ignored them, stared out the window and wished he were anywhere else.

Bill was nice enough, and that was the problem, really. He was too nice. He didn't bother to say anything to the kids and they took that for permission to do anything they wanted to do—which stopped just short of unscrewing each other's heads.

When they had driven fifty or sixty miles Bill suddenly slowed the van and took a narrow, winding road that led off east into the desert, toward the canyons.

It did not look like they were going toward any kind of campground, Brennan thought, looking over his mother's shoulder out the front window of the van. The road grew worse and worse and at last they were winding over rocks and sand and through sand dunes on little more than a narrow trail. Eventually even that ended.

"Now," Bill said, getting out and stretching, "we walk."

"Walk?"

Brennan couldn't help himself, his voice had a definite down tone to it and his mother shot him a warning look. Walk—with these kids? Brennan got out of the van and shook his head, you'd have to have a whip and a chair.

But if Bill heard his voice he gave no indication. He took gear from the rack on top of the van, laughing and talking and pointing.

"See that canyon?"

Over them rose the rock canyons leading up to the mountains. They had driven past several of them on the highway and turned into the fourth or fifth one. Brennan couldn't tell for sure now that they were so close to the bluff wall.

"It's called Horse Canyon. I came up here once years ago hiking and found a trail to a spring in the back of it. That's where we're going, back up in the canyon and camp by the spring."

Brennan looked up at the canyon. It seemed to be all rocks and cliffs—he could see no evidence of any kind of trail at all. The bed of the canyon, which lay straight before them, seemed to be an old riverbed filled with enormous boulders. It was impassable. To get "back up in" the canyon it would be necessary to work along the river and Brennan couldn't see a way to do it.

"I don't see a trail," he said, trying to keep his voice light. He still did not want to ruin this for his mother.

"Don't worry. There's one there," Bill said. He gave each boy a sleeping bag to carry and shouldered a pack, motioned Brennan to take another pack and Brennan's mother to pick up the rolled-up tents. "Really, it's not hard at all."

Which was not quite accurate.

It was true that it wasn't impossible, as Brennan had thought, but it was hard enough so that Brennan had to help some of the smaller children in a few of the places—just to climb over boulders and cuts across the trail—and even Brennan's mother felt it was a bit much.

"Do we have to go all the way to the rear of the canyon?" she asked at one point.

"Aren't there some nice places, you know, closer?"

But Bill insisted. They worked their way up on a small trail that led along the dry riverbed filled with boulders and smooth, dishlike bowls made by an ancient river that had roared down the canyon. In some places newer runoff had cut across what little trail there was, carrying it down into the riverbed, and here they had to drop into the cuts and climb out the other side.

Whatever else he is or isn't, Brennan thought, watching Bill up ahead with the pack on his back and helping the small boys—he isn't a wimp. Brennan was in good shape from running and he was soon breathing hard and his mother, who did not run or exercise very much, was almost staggering.

It was the sort of place Brennan would have loved, had he not been with the rat pack, as he thought of them by this time. The canyon walls rose straight up into the late afternoon sky; towering red and yellow with dark streaks, one bluff feeding to another up into the mountains. To say it was beautiful, he thought, seeing them shoot up over him, was just not enough. The beauty seemed to come almost from inside his mind, so that he saw the cliffs and canyon walls as if he had almost painted them. Here and there a scraggly pine hung on to life and yucca plants and cactus in bloom made color spots that seemed to make the cliffs even more striking. Art, he thought—it was like art. The year before, he'd taken an art appreciation class—largely because he had to—and the teacher, a short woman named Mrs. Dixon, had spent many periods trying to get them to see the art.

"See it," she would say to them, holding up a picture of a painting by Rembrandt or van Gogh or Whistler. "Really *see* it, inside it—see the brushstrokes? See what the artist was trying to do?"

And he tried that now, to really *see* the canyon, see what the artist was trying to do with it.

God, he thought—that was the artist. What was God trying with this?

Except that he was seeing the beauty with one eye while he was trying to watch ahead with the other and keep up with Bill and his mother and the kids, who were all over the place, sticking their fingers in cactus and screaming, throwing rocks down into the riverbed to hear them bounce, spitting off of boulders, hitting each other, jerking their pants down and mooning each other, throwing rocks at each other . . .

It was hard to see the beauty.

It took them almost two hours, until evening, to get up into a point where the canyon seemed to flatten out a bit and then another half hour of walking to get across a grassy area to a place where a trail dropped down into some small cottonwoods that were growing around a tiny pool of green, clear water.

"Oh," Brennan's mother said. "Isn't it pretty here?"

Bill turned and smiled at them. "It's like a calendar picture, isn't it?"

And it was almost sweet. Even the monsters stopped and were quiet for a moment or two. In the sudden silence Brennan heard birds singing and felt the sun on his neck. It had been hot but there was a coolness in the evening air that felt refreshing. Brennan lowered his pack and sleeping bag and two other bags and packs he'd been carrying for the younger boys.

"Is this it?" he asked, looking up and around. They were in the end of a box canyon and the walls rose above them making a huge amphitheater. The kids had been yelling and whistling for half an hour, listening to the echoes, but in their silence Brennan could hear himself breathe.

Without knowing quite why, he held his breath.

And in that instant something about the place took him, came into him and held his thoughts. Something he couldn't understand. Some pull, some reaching and pulling thing that made the hair stand up on the back of his neck.

He looked to see if any of the others felt it but they didn't seem to. Bill was still holding his arms out and smiling, the kids were beginning to move again, and his mother was in the act of dropping her pack on the ground and blowing hair out of her eyes.

He let his eyes move up and around the canyon again but could see nothing out of the ordinary. The rock cliffs towered over them, blocking the sun off so that now though it was still light they were in shade and it became almost cool.

Birds flew across the canyon. High overhead an eagle caught a thermal and wheeled out of sight past the edge of the canyon wall.

He shook his head. Strange, the feeling, he'd never had it before and now it was gone. Somebody was saying something and he looked down and saw Bill nodding.

"Yes. This is where we spend the night, where we camp. Let's get wood for a fire and make some dinner. I'm sure everyone is starved, aren't we, boys?"

The monsters started screaming and running in circles grabbing sticks and Brennan had to fight to keep from yelling at them.

It was going to be a long, looong night.

7

Visions

He did well.

During the night they went past a camp of bluebellies who had stopped on one of their patrols and Coyote Runs did well.

There were only eight of the bluebellies and they had a large fire as they always did to keep away the darkness so it was easy to see them and count them and would have been easy to kill them but Sancta shook his head.

Not this time. They stood for a time not a long bowshot from the soldiers and their fire, pinching their horses' muzzles so they would not make sounds to the soldiers' horses, so close the glow from the fire lit their faces, and could have gone in amongst them. Coyote Runs had his bow ready, an arrow on the string, and knew he could hit one and maybe another soldier as they rode in, knew it in his heart but Sancta said no.

Shaking his head once, a jerk from side to side that they could not miss and the old leader turned and led his horse back into the night.

And Coyote Runs brought his pony around and led it silently in back of Magpie and did not shoot an arrow at the camp, as he wished, but held back though his neck was stiff and swollen with the need to go amongst the soldiers.

They rode all night. There were no watering places for the horses but they had carried water in clay pots tied to their horses' blankets and stopped to walk and moisten the horses' mouths to keep them moving.

Twice Coyote Runs' pony seemed about to go down and he filled his mouth with water and spit it down the horse's throat so it would not waste out the side of the horse's mouth and the pony kept moving and he thought:

I am doing it. I am doing well. I am a man.

They rode all that night until the ridge led to a flat place with dead grass that seemed to go forever and here they stopped and made a cold camp.

They did not stay more than three hours, enough to let the horses rest but not so long they would stiffen up, then Sancta started walking again, leading his mount, and they all followed him.

Always south.

Sancta trotted ahead of them like a hurrying bear and after a time Coyote Runs realized he was beginning to see things. There was gray light coming from the east, over his left shoulder, the beginning of the new day.

Ahead he saw Sancta rolling along, his shoulders moving from side to side, his knee-high moccasins kicking up dust as he ran, his long black hair swaying from side to side and Coyote Runs knew he looked the same.

Knew that he could run this way all day, all night, as long as Sancta would need him to run.

Off to the right, many miles away, he could see the mountains that stood over the place the Mexicans called El Paso del Norte—the pass of the north—where the town lay, the white town and more, the fort where the bluebellies lived, Fort Bliss. That was there as well.

Coyote Runs had never been south this way and so had never seen the fort but he had heard about it. Buildings made of dirt as the Pueblos and Mexicans made them, arranged in a square for protection.

As if, he thought, as if the Apache would come against them in the fort. As if they would simply ride in against them in the fort and die.

Ho! There was madness there. In the bluebellies. The same as they wore dark wool clothes in the summer, buttoned tight to the collar so the heat could not get away and heavy hats to keep the heat in their heads and rode their horses through the heat of the day—all crazy.

But they fought well. He had never fought them but he heard stories, from Sancta and the others, sitting around the fire—stories of their fights with the bluebellies. They were not easy to kill.

It was hard light now, the sun showing, and he could tell that Sancta was looking for something as he ran, leading his horse.

In moments he swung off to the left and brought the party to the edge of a depression in the desert, a low gully surrounded by mesquite that hung out over the edges. It was as far across as Coyote Runs could throw a rock and Sancta led them down into the bottom of it.

"For the day," he said to them. "We stop here. Where we can't be seen by the bluebelly patrols. Sleep."

Coyote Runs led his pony beneath an overhang of mesquite so there was some small shade and lay on his back on the sand with the rein to the bridle tightly wrapped around his hand.

Magpie settled in next to him so the two horses stood closely together and could use their tails to switch flies from one another.

"Well, how do you like your first raid so far?" he asked.

But Coyote Runs was already asleep.

Brennan had thought as soon as it was dark the children would settle down, or at least slow a bit in their behavior.

He was wrong.

If anything they became more hyper and while Bill —Brennan was starting to think perhaps he wasn't really aware of things, like the world around him—while Bill took it well and no matter what they did seemed largely to ignore any of the problems, Brennan quickly became sick of them.

Short of tying them down—and he wished he'd brought rope—he couldn't control them at all. They were in his pack, in his bag where he'd spread it by the fire pit, throwing his gear around.

But after a time he noticed that as darkness came into the canyon, dropping like a black sheet down over them, the boys stayed more in the area of the fire. They didn't act afraid, but they didn't want to leave the glow of the fire either and that gave Brennan an idea.

He moved his sleeping bag farther and farther from the fire until he was nestled beneath a large boulder sticking up out of the ground, back under the overhang. There was a small sheltered area here, only large enough for one bag, and he spread it out on the dried grass and began to make himself comfortable.

"What are you doing?"

His mother had walked up in back of him while he was spreading his bag.

"Don't you want to be around the fire?"

"Mom . . ." He started to say something about the kids but let it go. She was so happy, or seemed to be happy with this Bill character, that he didn't want to do anything to ruin it for her. "This just looked like a neat place to set up."

She nodded—the glow from the fire lighted her cheek and it looked golden. Something about her face looked young, very young. "I know why you're up here—they're horrible, aren't they?"

"God," he said. "Like animals."

"Maybe they'll sleep pretty soon."

Brennan smiled. "Not unless you drive wooden stakes through their hearts. . . ."

She laughed and he'd somehow never felt closer to her. "But he's nice, isn't he? Bill, I mean."

Brennan nodded. "Yes. He is. I can't believe what he takes from them and never gets mad."

"I like him. A lot."

"That's nice—really."

"A lot."

She turned and walked back to the fire and he watched her go and felt a kind of sadness. There had been others that she had liked. A lot. And they somehow all came to nothing and she wanted somebody so badly, needed someone so badly that Brennan almost wept for her sometimes.

He rose from the bag and moved toward the fire. What the heck, maybe if he took the kids for a while Bill and his mother would get a little time to know each other. What would it hurt?

He sat by the fire. "Who knows a story?"

And it worked. They all settled in around him, sitting by the fire in the yellow glow, their faces looking up at him. It couldn't be, he thought, it couldn't be the same group of wild things that just a few minutes ago were tearing each other apart.

"I said," he repeated, "who knows a story?"

They all looked at each other and shook their heads. None of them knew a story.

Bill and Brennan's mother sat next to each other across the fire from Brennan. Bill coughed.

"I know a story," he said, "or not a story so much as some information about these canyons."

Brennan could see that it wasn't exactly what the kids were expecting, but they held still and waited.

"There are four or five of these canyons along this big ridge," Bill said, pointing up over his shoulder where the cliffs rose into the mountains. "And some called it the last stronghold of the Apache nation."

That got them. It also got Brennan. "What do you mean?"

"I'm kind of fuzzy on it, but north of here there is a canyon called Dog Canyon that also has a spring in it. I guess the Apaches would raid down south and then run back to these canyons to hide and the army would come after them."

"Were there battles?" one of the small kids asked. "With shooting and blood and guts and stuff?"

Bill nodded.

"All *right*!"

"There were several fights in Dog Canyon and there might have been some others in this canyon, in all the canyons along this ridge line. Soldiers and Apaches were both killed and . . ." here he paused and looked at Brennan and hid a wink ". . . they say that in the night sometimes you can hear the sounds of battle and that the ghosts of the dead warriors and soldiers walk in the darkness."

He stopped talking and there was silence around the fire. Far away a nightbird called and the sound seemed almost human—close enough so the boys drew together.

"I don't think I would run around too much at

night," Brennan said, taking advantage of the situation. "You know, out away from the fire."

But he hadn't needed to say it. Two of the boys, a set of twins named Glockens, said they weren't afraid of any old ghosts but they didn't stray more than ten feet from the fire and were very happy to toast marshmallows and drink hot chocolate made by Brennan's mother and Bill.

Bill told some more stories—plain ghost stories—and the children sat still, listening, and finally it was time for bed.

Brennan nodded good night to his mother and Bill and headed up to his bed beneath the rock, carrying a cup of hot chocolate. There was a light dew condensing on his bag and he shook it off before crawling inside. Then he sat up, the bag around him, and looked down at the fire below as the rest of the group crawled inside their bags, sipping his hot chocolate.

It wasn't so bad, this trip, he thought. The kids had finally settled down and his mother was right, Bill was a good guy. He hoped well for his mother.

The fire died quickly and he could hear the kids mumbling and talking for a short time, then it was quiet and still he sat.

He felt strange. The chocolate grew cold in his hand and he sat and let the canyon come in around him. There was no moon, but enough light came from the stars so that when his eyes grew used to the darkness he could see the canyon walls moving up into the sky.

Somewhere far away something screamed a faint cry

—almost like a woman or child screaming—and he started, then remembered reading somewhere that mountain lions screamed that way and thought it must be a cougar somewhere way off. There were mountains and more mountains up over the ridges and there must be mountain lions up there in the peaks.

I have lived so close to this, he thought, and never been here, never seen this beautiful place. He had been in the mountains close to El Paso but they were dry, dead, hot and baked airless peaks—not like this. Not with cool breezes and springs and cottonwoods and birds and lions, if it was a lion. He wondered for a moment if he should be frightened of the lion, then decided against it. From what he'd read they didn't bother people at all and it must have been miles away. The sleepers below him hadn't even heard it.

He put the cup down and lay back but still sleep wouldn't come.

Something was there, some strange thing that bothered him. He had felt it before when they first came into the end of the canyon and it was still there, the feeling. He couldn't shake it.

An unease, a restlessness that wouldn't go away. He closed his eyes and thought of things to make him sleep, boring things, but even that didn't work. In the end he sat up again, staring out across the canyons over the sleepers below him, a strange uneasiness in his heart that would not go away.

And the night came down.

9

The Raid

Oh yes, it had been something, the raid, Coyote Runs thought, holding tight to the pony with his legs. It had been a thing to see.

They had ridden through part of the night, spitting water into their horses' mouths until they arrived at the river where the horses could drink, just at dawn, and had found the large horse herd before the sun had risen completely above the line of land to the east.

Sancta had motioned to Coyote Runs to ride up to him and he had done so.

"Stay here with the extra horses. We will ride amongst them and take many horses and come back this way. Let us go by, then come in at the rear of the herd to help move them."

Coyote Runs wanted to shake his head, wanted

to say that he wished to ride amongst the Mexicans who were watching the horses to prove that he was a man, ride amongst them and fight but he did not. Sancta had given an order and that was the way things would be.

But he could see the horse herd, or part of it. Sancta and the others left him on the back side of a small rise but as soon as they rode off toward the Mexicans Coyote Runs moved forward and up onto the rise slightly so he could see some of it.

There were so many horses that they stretched as far as he could see into the mesquite and gullies. Many, most of them, were large and brown, as the bluebellies liked them, but many were of other colors and Coyote Runs saw a large horse the color of straw that he would have liked to take for his own. Indeed, his hands moved on the reins of the pony and his knees closed before he remembered that he must stay.

He saw one Mexican rider, a tiny figure on the other side of the herd, but the rider did not see him and in any event it made no difference. At that moment when he saw the Mexican rider there was a commotion to the right and a group of horses that had been down in a low swale suddenly broke into a run, heading straight for the river.

For a moment Coyote Runs could see no Apaches, then he saw heads above the dust and knew that Sancta and the others had cut the smaller herd away from the main body and were bringing them north.

They were not coming straight toward him but off to the side a bit, so he yelled and slapped the horses around him to get them moving to meet the others.

Dust and noise were everywhere. There was little wind in the morning, so when the horses raised dust it simply hung where it came up, and was added to by other dust until it was impossible to see anything and the hooves of the running horses—Coyote Runs thought that Sancta and the others must have cut out over a hundred of them—were like thunder.

Coyote Runs was confused for a time and actually drove the waiting horses that he was in charge of the wrong way, headed them south. In the dirt and noise it was easy to be wrong.

And there he could have died, he found, because he rode straight into a group of four Mexican riders. His small herd turned and started north, going right around him, as if sensing they were going the wrong way when they saw the Mexican vaqueros.

It was impossible to tell who was more surprised. He pulled his pony up, staring at them, and they pulled up, staring at him. Then two of them started shooting at him with their horse pistols that made clouds of smoke to mix with the dust and he wheeled his brown pony and followed the herd back to the north.

There was no sense to anything.

Because he had initially wound up going the wrong way, when he started back to the north he found himself to the rear of the main body of Mexican riders who were chasing Sancta and the others toward the river.

He did not know this at once, but suddenly found himself riding next to a Mexican man who looked sideways at him, drew a pistol from his belt, and aimed at Coyote Runs and shot.

Coyote Runs winced, waiting for the bullet to take him, but the Mexican did not aim well and it went wide and he veered away in the dirt clouds.

He passed others, driving the extra mounts north, passed two more and then three, all riding hard after Sancta but none of them shot at him and he thought he must have been given special power, special medicine to go through them like smoke so they could not see him, though he rode right next to them and had horses in front of him.

Suddenly he found himself running in water, the little pony almost going down when it hit the river. It was impossible to see anything, to know anything, and he merely hung on and hoped he would make it across the river.

Still noise, the pounding of hooves, and dust, even over the water dust that blew off the banks, but he believed in himself now, and his new medicine, so he drummed his heels into the brown pony and screamed at the horses in front of him and drove them across the water and up the bank and kept driving them.

North.

If he kept pushing them north he would catch up to Sancta and the others.

So he followed the dust and knew that he was a man now, knew that he was a man with large medicine who had passed the test, for had the Mexican not shot at him, shot directly where he was and the bullet had not hit him? And he had not been afraid. He had tightened his stomach to take the bullet, had waited for the shock of it, but had not been afraid and was not afraid now.

He drove them, slapping at the horses with the end of his bridle rope, keeping them driving in an easy lope ahead of him, following the dust for a mile, then another mile, and was on the edge of wondering if he would ever see the others when he ran into Magpie.

Who was aiming his old buffalo shooting rifle at him.

"Wait!" Coyote Runs said, pulling up. "It is me."

Magpie lowered the rifle and smiled. His face was so covered with dust that it seemed to crack when his lips moved. "Sancta heard hooves back here and thought it was some Mexicans still riding after us and sent me back to slow them down. It is lucky you called. I had begun to tighten the trigger."

It would not have mattered, Coyote Runs thought, because the bullet would not have been able to get through my new medicine, the medicine that protects me, but he said nothing.

"How far are they ahead?" he asked. He must look as Magpie did, with dust thick on his face and sweat streaks cutting through it, but the band of red cloth around his head kept the sweat from his eyes.

"They are by now another mile," Magpie said. "Come, let us catch up to them before some more Mexicans really do come and we are forced to fight and kill many of them and make their women sad."

Magpie swung in beside Coyote Runs and they pushed the spare horses into a run again, although not as hard as before. They did not really think the Mexicans would come because they did not like to ride across the river where they might run into the bluebellies. It was a fact that at times the bluebellies did not care what they shot as long as they shot something and the Mexicans did not like to face them.

They rode that way for a mile, then another half a mile, eating the dust from the horses Sancta and the others had taken, riding in silence as they pushed the horses ahead of them.

They caught up as Sancta was pushing the main herd of stolen horses across a dry lake. A small wind had come up and the dust blew away so Coyote Runs could see the horses ahead and below on the lake and he drew in his breath.

Such a herd he had never seen or thought to see. The village normally had a herd of thirty or forty horses, which included all the horses of all the men, but Sancta and the others were pushing well over a hundred horses ahead of them. Enormous wealth.

It was a great raid, a raid they would speak of for years around the fire and he, Coyote Runs, was part of it.

And now a man.

Magpie looked at him and made a sound in his throat, an almost growl of pride and exultation, and Coyote Runs matched the sound with one of his own.

"We have ridden amongst them," Magpie cried, his voice raw and low from the dust, "and taken many horses! We are the ones, we are the ones!"

They rode off down into the lake bed after Sancta and the rest of the men and horses with wild cries. Even the ponies seemed to have caught the excitement and Coyote Runs had to hold back on the rope to his pony's halter to keep him from running himself out.

They caught up with the main herd at the edge of the lake and there was much laughing and joking.

Not a man nor a horse had been hit, though the Mexicans had fired many shots at them, and Coyote Runs thought it must be that his medicine was so strong it had extended to cover all of them.

Truly, he thought, truly my medicine is strong and he wondered how that could be since he was so young. Most strong medicine went to older men. When they got back he would have to ask his mother about his medicine dream, his name dream. Was it stronger than other men dreamed? He had never heard such a thing. But his mother knew of dreams and visions and would be able to tell him.

Magpie and Coyote Runs put their small herd in with the main herd and they all started north. The men were positioned all around the horses, with Magpie and Coyote Runs bringing up the rear since they were the youngest and as such would get the least favored position. There was even more dust and dirt than before and soon Magpie untied his headband and tied it around his nose and mouth to filter out some of the dust.

Coyote Runs did the same and they rode that way all of the day without a single stop, until close to dark, when Sancta called a halt to rest the horses and change mounts.

They did not build fires.

"There may be riders coming after us," Sancta said, "although I do not think so. It will be better not to give them fires for guidance. Drink only a small amount of water and spit some into your horse's mouth. Take a Mexican horse for a new mount, they are still fresh."

Coyote Runs had his eye on a small reddish

horse and was making for it when he saw a flash of color out of the corner of his eye and turned to see the straw-colored horse standing sideways to him. It must have broken from the rest back before the river and come on its own.

He was the youngest and had to wait for the rest of the men to get horses. But none took the straw horse and when all had new horses he rode up to it.

It did not run from him.

He slid from Magpie's brown pony, patted it on the neck—it had run well for days without faltering—and slipped the end of the bridle rope from its jaw and onto the jaw of the straw-colored horse.

It stood for it and kept standing while he tied the blanket on its back and moved his water bottle and bow case. The brown pony moved in with the rest of the herd and he grabbed the long mane of the straw-colored horse and threw his leg over its back and settled onto him.

The horse acted as if they had always been together. He answered knee pressure, wheeled around, back and forth—Coyote Runs smiled, then laughed.

"You have a horse," Magpie said, riding up alongside him on a black mare with a white blaze on its forehead.

"Truly," he answered. "I hope I can keep him."

"Do not worry. You will get him. You have

done well on the raid and Sancta will give you the horse. I know it."

Coyote Runs hoped he was right but said nothing. It would not be proper.

They started north again, but now that they were around the immediate area of Fort Bliss and the bluebellies Sancta cut the herd back toward the west a bit and headed northwest, not back toward the village but off to the west of it. Riding this way they would not come back to the top of the canyons at all but out in the desert in front of them.

Coyote Runs moved over to Magpie and rode next to him for a time.

"Why is it that we do not head back for the village?" he asked.

Magpie snorted through the cloth over his mouth and nose. "You have much to learn still. You never come home directly from a raid, never show where the village is. Sancta leads us this way in case anybody follows us. We do not want to lead them back to the women and children, do we?"

It made sense and Coyote Runs wished he had kept his mouth shut. He moved the straw-colored horse back to the side and thought how stupid he was—a true warrior would have known, would not have asked, and he vowed never to ask anything stupid again.

Dust.

It grew thicker as they rode in the dark until

Coyote Runs had no idea of where he was, where he was going or where he had been. He could not see the stars, the horses in front of him—he could not even see the ears on the straw-colored horse he was riding.

All night in this manner they rode. Not fast, because it was not possible to push the large herd with any speed. If pressed, the herd would just break into smaller groups and scatter and they would lose many of them.

But they kept up a steady movement through the night—although Magpie and Coyote Runs had no idea how fast or how far they had gone—and by the time first good light came Coyote Runs was all but falling off the straw-colored horse.

Out of excitement he had slept poorly the day before and now a whole night with no rest had added to the exhaustion and brought him to the point of reeling.

He took some heart in the fact that Magpie was in no better shape. At one point he found himself riding next to his friend and saw that he had tied himself to his blanket cinch and was sound asleep, his head nodding forward on his chest.

Coyote Runs took a small cord from his bowcase and did the same thing, going around his waist and to the rope that kept his blanket in place. He was just pulling the knots tight when he heard gunfire.

In the thick clouds of dust rising in the morning air it was impossible to tell what was happening but he could hear the fire and knew whoever was shooting had large guns. The sounds were low, thudding, and the Mexican riders would not have such guns. Only the bluebellies, the pony soldiers, would make such a noise with their large rifles.

He hesitated for a moment, confused, and in that time Magpie—who had been on his left—came galloping out of the dust.

"Run! Soldiers—run, my friend, they are too many. They are amongst us. Head for the canyons. Run now!"

And he was gone, off to the right in the dust, invisible.

Still Coyote Runs held back. He was reluctant to lose the horses so easily. What would the men say if he just ran and it proved to be nothing? Besides, his medicine was strong. Had not the Mexicans shot at him and missed?

But in half a second another figure came out of the dust, then two, and he saw that they were wearing the blue wool coats of the soldiers and were holding the loud rifles and were aiming at him and he turned and dug his heels into the straw-colored horse.

It broke into a run as if waiting for the command, lunged so hard that Coyote Runs would have fallen off had he not been tied on. And the lunge saved his life, as he fell backward a soldier rode beside him and held out his rifle and fired, not ten feet away, but the bullet passed where Coyote Runs had been sitting and missed him.

Now run, he thought. Now little horse, run for all there is, run for my medicine, my life, my soul. Run like the wings of birds. Fly—runflyrunfly.

The straw-colored horse laid its ears back and streamed its tail and streaked through the sand dunes and mesquite so fast that the soldiers could not possibly have followed him, would have lost him in the thick clouds of dust.

But suddenly they broke clear into the morning sun. Running across the rear of the herd Magpie and Coyote Runs quickly moved out of the dust cloud and as they did Coyote Runs recognized where they were.

He was running straight toward the bluffs and high canyons up from the desert. Ahead of him not two hundred long paces he saw Magpie driving his horse, whipping it, and when he turned to look back he saw four soldiers break out of the dust, chasing them. Three hundred paces, no more, separated him from the four men. They all had the large rifles. One of them had been the one to shoot at him.

Four, he thought—so much noise from four men?

But there were more. Off to the side, looking to the north, he saw the main body of Apaches were being chased toward the next canyon over by more soldiers.

Firing.

Everybody was firing. The soldiers were trying to shoot while they rode. It was hard to see how they could hit anything, but Coyote Runs heard those behind him fire—the rifles making a dull thud—and then heard the whistle of the bullet passing close to him.

The canyon. How far?

Bright sun, clear morning desert air, cool morning air with the horse running so well, how far? How far to safety?

Ten, fifteen bowshots to the mouth of the canyon. And then what? If he made it, then what?

His medicine. The canyon they were heading for was the one leading to his medicine place. If he could make that, could get to the sacred place of the ancient ones, surely he would be safe. . . .

More firing, the bullets hissing past. He turned to look back once more and saw that the soldiers had fallen slightly back. They were all big men, heavy, and their mounts were not as fresh. It must have been the group of soldiers they saw on the way down—out on patrol. They must have run into them. He saw sweat on the sides of their horses, foam from their mouths. His own horse was the same, wet, but still driving well, its shoulders pumping with great strength.

O spirit, what have you given me here, he thought. What a horse.

Four bowshots left to the canyon mouth.

There he would have to leave the horse and run

up the sides on foot. If he got into the rocks they would never catch him. Not the big sweaty men in the heavy blue clothes—he would be free.

Again they fired. Again the small whistles, the little chu-chu-chu of death.

Ahead of him Magpie suddenly jerked erect on his horse and a red spray went out from his chest. He threw up his arms almost as if he were waving both hands. He began to fall forward, then fell back and to the side and was dragged by the rope around his waist. His horse tried to keep running without stepping on him but could not and veered off into a circle fighting away from the body.

The body.

Magpie.

Magpie was a body.

He must untie the rope holding him to the straw-colored horse. He fumbled with the knot, jerked, finally pulled it free. Only twenty good leaps to the canyon mouth, to the trail, to safety, to life.

Magpie was a body.

He slammed his heels into the horse, asking for more, still more speed. Behind him there was more gunfire. The bluebellies reloaded as they rode, shot again and again and still the bullets missed.

Ten leaps.

Now five.

Coyote Runs felt a slap on his leg and the horse

grunted beneath him and began to go down. They had hit the straw-colored horse. But when he looked down to the side he saw his foot hanging loosely, blood coming from just above the ankle.

They had hit him and the bullet had gone on into the horse. His medicine had failed—how could that be? He was so sure of it. . . .

No thinking now, too late for anything.

The horse collapsed, its legs getting softer and softer as it caved in and just as it hit the ground Coyote Runs fell off to the right and rolled on his shoulder. His bow, all his arrows were gone. He had no way to fight but it didn't matter now.

He was in the canyon.

The mouth of the canyon.

His medicine place—he had to reach it. It was all he could think of now, pulling himself along, and he scrabbled on his one good leg and his hands up a narrow trail, kept going though the pain came now in waves, covered him in red waves, kept pulling and fighting until he was in a grassy area.

They would not come, he thought. The soldiers would not come after him up in the canyon. He would keep going but they would not come for him. They would turn away.

He was wrong.

He heard them yelling in back of him, yelling to each other as they started up the trail after him. Their voices echoed from the canyon walls. They did not sound like men, the voices, but like devil voices, death voices, ghost voices.

He shook his head. The craziness from the wound was coming into his head and he shook it to clear it, to stop the weakness.

He needed everything now. Had to have everything in him to get away.

He fought forward on his good leg and hands, crawling and hobbling across the grassy area, the dry grass crumbling beneath him, the morning sun warm on his back.

Everything bright, everything very clear and bright and hot and fresh. The air smelled good, even through the pain; smelled sweet and good.

I will do this thing, he thought, his head momentarily clear. I will get away from the bluebellies and to the place of the sacred ones and back to my mother and ask the question of medicine, ask what I do not know.

The soldiers voices grew fainter.

I am doing this—I am making it away from them. O spirits, dust spirits and wind spirits and ghost spirits help me, come now to help me.

They were in the streambed of the canyon, down and to his right, between him and the place of the ancient ones. They blocked him. He would have to hide. He would have to find a place and hide from them and let them search until he could get around them and while thinking of it, while wondering where he could hide, his eyes caught the dark place beneath an overhanging boulder looking down on the spring.

All at the same time he saw the spring and the boulder and the dark place and knew what he had to do. His body, his whole being wanted to get to the water at the spring. He had never been so thirsty. But that was where the soldiers would look. They would seek him there.

Instead he must hide beneath the boulder. Cover himself with dirt and hide there and let the spirits take the soldiers the wrong way.

He worked in beneath the boulder, back in the crack where it met the earth, and carefully covered himself with sand and dust so that he would not show and thought still that he would do this, could do this thing.

And now they came.

Three of them. They must have left one to stay with the horses. Three of the soldiers came and he smelled them before he saw them. They smelled of strange sweat and some smell that came from the wool in their clothing and tobacco and hair on their faces—some mixed small of all that together.

White smell.

Bluebelly smell.

Pony soldier smell.

Death smell.

Then he saw them. All three were walking side by side, about ten yards apart, staring intently at the spring, the small cottonwoods, the big rifles held in front of them, ready.

Ready for him. Ready to shoot him.

Now, he thought—now must my medicine be strong and the spirits come to help me. Now there must be help.

They were so close now, so close he could see that one of them had a small cut on his cheek and that the blood had dried black. The man was large, squinting in the sun, and in the same sun, in the new morning light Coyote Runs saw his end, his death.

As he watched the soldiers begin to pass, his eyes fell on the ground in front of them and there it was, his betrayal.

His ankle had left a small trail of blood, smears here and there on the rocks and in the dirt. He had not thought of that. Had not considered that he would leave blood.

Still he had a moment of hope. They were almost past it, past the small blood trail and he thought, oh, yes, I will live yet, I will not die in this place. I will live, I will live, I will live. . . .

Then the bluebelly saw it.

The soldier on the right stopped suddenly and

looked down and Coyote Runs thought, no, not now, hide me medicine but knew, knew it was too late.

The soldier's eyes followed the scuffs and patches of blood up and to the right, up and up the eye came until he was looking at the rock.

He said something, a low word to the other two men, and they stopped.

Take me now, spirit, Coyote Runs thought—take me up and away now, away and away from this place. Take me.

They saw him.

But they did not shoot.

They walked up to where he lay beneath the overhang of the rock, stood there, strong and tall and ugly and blue and stinking of white sweat they stood there and looked down on him.

Coyote Runs did not move, lay looking up at them, sought his spirit, sought his soul. Away now, take me away from this place, spirit. . . .

The man who had first seen him said something again, not to Coyote Runs but the other two men, and they all laughed. Words Coyote Runs did not understand, but the laughter he knew. It was hard laughter.

Then the soldier said something to him.

It was some kind of order, but he could not understand it. Some strong word. Maybe he swore.

Coyote Runs looked at him, shook his head. I do not know what you mean.

Then the soldier leaned down, still smiling, and put the muzzle of his rifle against Coyote Runs' forehead and he thought, no, not now, I will go with you, I am in the wrong place, take me, spirit, take me now quickly before, before, before. . . .

There was an enormous white flash, a splattering flash of white and the start of some mad noise to end all noises and then there was nothing.

Nothingness.

10

Brennan snapped awake and sat up.

He was scared, no, worse, terrified. Short breaths, pants that puffed in and out, his eyes wide, nostrils flared and the night, the night all around and closing on him and he did not know why.

Did he dream?

He could not think wholly of it, if there had been one. There were the edges of something, some greater horror that he did not understand. A stink he did not know, and pain, horrible pain in his leg and head and it had slammed him awake.

Hard awake.

He fought to control his breathing, got it down, stared around in the darkness. There was only starlight but it made the canyon seem painted with a silver ghost brush. He could see things, boulders, canyon walls, but not in detail. As if painted and left blurry.

Take me, spirit . . .

The thought came and left before he knew he'd really had it. Just the words. *Take me, spirit . . .* in and out.

How strange. I never think these things, he thought. What is wrong with me?

Below him the others slept quietly. The fire had long since died out and even the ashes must be cold, he thought, and the word *cold* brought the night down on, into him. He was suddenly freezing, the sweat chilling him because he was sitting.

He pulled the bag up around his shoulders and shivered. By squinting he could just make out the numbers on his digital watch.

Three in the morning.

He had never been a light sleeper. Even just after his father left when his mother sat crying Brennan would sleep hard. And he never, ever had come awake suddenly so frightened, his breathing stopped.

Or thinking strange thoughts about spirits.

He lay back and pulled the bag up over his head to hold the warmth. Even in the summer, in the desert the nights were cold.

What is the matter with me? Nightmares, cold sweats.

He tried to close his eyes and doze but could not sleep. There was great unease in him, a great churning, whirling confusion that he could not understand and he sat up again.

Something.

Something was there with him, in him, around him. He could feel it. Some other thing was there.

What are you?

It finally came, the question. He asked it in his mind, then aloud, in a soft whisper:

"What are you?"

But there was no answer, no sound but a nightbird off in the canyon moonlight. A low warble, soft, there and gone. Below, his mother must have heard the sound in her sleep and turned over in her bag but did not awaken.

He was still very cold, colder than he'd ever been. He wrapped in the sleeping bag and shivered but could not seem to get warm and when he lay back on his foam sleeping pad he realized that something was sticking up beneath his shoulder, pushing into the foam.

He tossed and turned but it still seemed to be prodding him and finally in irritation he raised up the corner of the foam pad to pull it away.

It was a round and fairly smooth rock, the curve of it sticking just out of the sand beneath the sleeping pad, but when he tried to dislodge it he found that it was buried and would not come out.

He pushed and pulled at it and after some effort it began to wiggle slightly and he at last broke it loose and put it over to the side. He scraped some sand into the hole left by the rock and lay back again but in all this effort the fear, the sweat, the chills had not left him.

Take me, spirit . . .

Again it was there, or still. Who are you? he thought. Then whispered it. "Who *are* you?"

And thought, I am cracking, completely cracking up. This is crazy.

Maybe it's where I am. Maybe there is something in the wind. . . .

There was no wind. The still canyon in the moonlight seemed almost to mock him. Everything was so peaceful, so settled and beautiful, and yet there was something that would not leave him alone.

He could not sleep and again he sat up and as he did so his hand fell on the round rock he'd dug out from beneath the sleeping pad.

He started to throw it away to the side but as he made the move the light from the moon caught it and he saw it for the first time.

"What . . . ?"

It was not a rock.

It was a human skull.

He dropped it instantly, jerked his hand away and pushed back under the rock. But there it was again, a soft nudge in his thinking, a command, no, a request, a thought to do a thing he did not understand and he picked the skull up again.

It was very old. Or he thought it was very old. And

it was not complete. There was no lower jaw and it was filled with dirt and small rocks so that he could not make out the features.

He held it up in the moonlight and turned it this way and that but there wasn't enough light to see very well. He shook it to knock some of the dirt out of it and there was a small rattling and all the dirt, dried old sand, fell away.

Suddenly there were eye sockets and a hole where the nose had been and teeth, upper teeth, but the whole back of the skull was missing. There was a gaping hole where the rear curve had been and when Brennan turned it over he saw that there was a round hole in the front of the skull, just above the eyes, roughly the size of his index finger.

A bullet hole.

It was the first thing that came to mind and he should drop it, he knew, should put it down. Whoever had been the skull, whoever it was had died a violent death, a murder victim, and he was playing with evidence and yet he could not.

Could not put it down.

Take me, spirit . . .

There it was again, while he held the skull. It did not come from the skull but from some other place, outside his thinking and yet inside his mind. Some strange part of him.

And he then knew what he had to do, knew it with all the certainty he had ever felt about anything in his life. He must take the skull.

Wrong or right it was there. A fact. A known. He must take the skull from this place.

He looked below to the fire pit and they were still all asleep, even the kids. But the sky was beginning to lighten over the back of the canyon, to the east, and they would be waking soon.

He pulled his pack close to the back and opened the top compartment. In the bottom he had a spare sweatshirt and he pulled it out now and wrapped the skull in it and put it back carefully and thought all the time, all the time, I am crazy, this is crazy, I am crazy.

But he could not stop himself.

11

Brennan shook his head and looked up just as the mower began to head into a rosebush. He jerked the wheel hard sideways and slewed around the edge, so close, the rose blooms brushed his cheek as he went by.

Nothing, no part of his life was acting normally since the camping trip and the skull.

A week had passed. They had broken camp after breakfast the next morning and walked back out to the van, the kids wild all the way, and driven home and nobody knew of the skull.

When they returned home Brennan put the skull up in his closet on a shelf reserved for models, wrapped in paper. His mother rarely came in his room and never bothered the shelf, so it would be safe.

Although safe from what, safe from whom Brennan couldn't think.

He was acting ridiculous and he knew it. One side of him told him to call the police and take the skull to them and tell where he got it. The skull might not be as old as all that, might be part of a murder investigation and he would be withholding evidence. He did not know how much of a crime that was but suspected it was a serious one and he was not a criminal. Or at least had never been one.

Before now.

Crazy.

But something in him, some other part of his thinking told him to keep the skull and he could not think of a single reason why.

To know it.

That thought crept in and had been in his mind before. Every time he thought of the skull. He was supposed to know it.

Take me, spirit had been there, in his thinking, and now, *to know it.*

Know *what?*

It was so frustrating! Here he was, hiding a skull with a bullet hole in it in his closet, hiding it even from his mother, maybe breaking the law and he had no reason for it, no excuse.

Well, see, Officer, every time I thought about calling you this little thought came into my head and it said something about a spirit and that I was to know something. . . .

He wheeled the mower around another rosebush and

looked up to see Stoney glowering at him from where he was working with the string cutter around the small flower beds.

The house belonged to a judge and Brennan kept finding himself looking at the windows thinking what if the judge was home and knew he had a criminal working in his front yard.

Possible criminal, he thought, make that a possible criminal.

The week had been almost completely insane.

He no longer slept well. At night in his room he lay with his eyes open, thinking of the canyon and the moonlight and, of course, the skull.

Always the skull.

The first night back he had taken it from the closet and examined it in better light at the small desk-table in his room after his mother was asleep.

It was so small.

That was what hit him the first time. The back half of the skull was gone but even if it had been there it was a very small skull and he thought it must have been a child's skull.

So he took a tape ruler out of the drawer and measured it, across and from front to back, as best he could and then—because there was nothing else to compare it to—he measured his own skull.

They were nearly the same.

He had to press through his hair to get the tape tight down on his own skin, and make a guess at the measurement around the missing back part of the skull but when he did there was only a slight difference in the measurements and he thought then that it must have been a boy.

A boy like him.

There was no reason to think it. He would not have a much bigger skull when he was a grown man, so why could it not be a man's skull? Or a woman's or girl's skull?

Yet he could not shake the feeling and that night, the first night back, he put the skull back in the closet, went to bed but could not sleep though he was tired from the camping trip and not sleeping well the night before.

At last his eyes had closed and he felt that he was still awake but somehow he dreamed, slipping in and out of the dream.

The dream that night made no sense. He sat cross-legged on a high ridge overlooking the desert and the canyons below, apparently near where they had camped, and watched an eagle flying. It moved in huge circles, taking the light wind, climbing and falling, and he just sat and watched it fly and didn't think or say anything, didn't do anything.

He could sometimes see the eagle very closely, see the feathers, the clear golden eye, then it would swing away and go higher and higher and finally become a small speck against the blue sky and then, in the end, nothing, and he just sat all the time on the ridge watching.

Then he had opened his eyes and was awake. When he looked out the window it was light, well past dawn, and he was surprised to see the clock on his desk at seven-thirty. Stoney was due shortly and he had jumped out of bed and gone to work without eating.

All that day, and the next, and the next he had thought of the skull. And during the nights he did not sleep but lay back and closed his eyes and had strange dreams until he opened his eyes and had to go to work.

After the eagle dream he dreamed of a snake. The snake was a rattler, coiled in a lazy *S* on a rock in the sun. Brennan was afraid of snakes, snakes and spiders, but in the dream he had no fear of the snake and sat near it on a stone and watched it. The snake moved this way and that, back and forth, but not forward, the head weaving gently, the tongue flicking out in silence. Then it swung its head around and looked directly at Brennan, into his eyes, and Brennan was still not afraid. In the dream he studied the snake and the snake seemed to nod, its head moving up and down once gently, and Brennan answered the nod and was awake and the dream made no sense.

He knew nothing of snakes. Except for one dead on the highway as they drove by he had never seen a rattlesnake, knew nothing about them. But the snake seemed to know him, seemed to be trying to say something to him, and in the same dream he danced with a group of dancers in a circle, holding arms and moving to the rhythm of a drum.

In the dream he looked down and watched his feet move in the sand, kicking up small puffs of dust, and he yelled with the rest of the dancers, yelled a word he did not understand when the dance was done.

And came awake. That night he sat up in his room for close to two hours after the dream and thought he must be going crazy.

Of course he knew it came from the skull, and thought in some way that it had to do with a guilty conscience that came of not calling the police, or telling his mother. That feeling guilty was making him have bad, or at least strange, dreams.

But when he came to the edge of it the next day, telling his mother or calling the police, when the pressure grew and bothered him that much he could not do it.

It was not that he didn't want to, not that he didn't feel like it.

He couldn't.

He simply could not make himself do it and that frightened him more than having a guilty conscience. Why could he not do it?

In the third dream there was a horse.

He was riding the horse in bright sunlight and heat. The horse had light, almost yellow hair but he thought of it as having the color of straw in the dream, a straw-colored horse, and it ran beneath him, between his legs with a power that seemed to come from thunder.

The front legs pounded up and down and he felt the back legs bunch and spring each time, driving the horse forward, the front shoulders rippling against his legs, driving, pounding.

There was great joy in the run, the wind against his face, the heat on his back, which was bare, and the horse thundering across the sand and his hair blowing out in back of him, his hair blowing like the straw-colored mane of the horse. He laughed in his throat in the dream and laughed in his bed and the sound awakened him and he came awake and sat up and was glad.

Glad. His heart felt gladness and he did not know or understand why. It faded, slipped away as he sat in the dark looking at the small light on the smoke detector thinking again, or still, that he was going crazy.

On the second to the last morning, two days before the day when he would come close to wiping out the rosebushes at the judge's house, he sat at breakfast with his mother.

She had been drinking coffee and he had a bowl of oat bran.

She was sitting in silence, sipping the coffee, looking out the window at the morning sun coming in, lost in thought, and he coughed to get her attention.

"Is it possible for crazy people to know they're going crazy?"

She studied him for a time over her cup. The famous mother look, as he thought of it. The Mother Look. The what-are-you-up-to Mother Look.

"That," she said, "is a very strange question for dawn on a Friday morning."

He didn't answer at once, thinking. "It's not me, understand. I was just wondering, you know, if someone is crazy do they know they're crazy?"

She put her cup down. "I don't think so—but I'm no expert. You'd have to ask a psychologist to be sure. . . ."

They had dropped it, and that night he had gone to bed half afraid to sleep.

And a dream had come.

This time there was another person in the dream. A girl. He could not quite see her as she moved ahead of him, walking somewhere—he did not know from where or to where—but he wanted to know her better. Wanted badly to know her. And she was gone, walking into a mist that he could not pass through, a mist that frightened him very much.

This time when he awakened he was drenched with perspiration and his mother was sitting on the side of his bed.

"You made noise," she said, "a funny sound, like words I couldn't understand. Do you feel all right?"

"A cold," he said, though he knew it wasn't true. "Maybe I'm getting a cold. . . ."

She had stayed with him until he had closed his eyes and feigned sleep but that night he did not sleep any more.

And the final day in the week he decided that he had to do something, find out what was wrong. . . .

Find out what the skull was doing to him.

And the best way to find that was to try to find out about the skull. He had to know more about it, all he could know about it.

That's where the answer was, somewhere in the skull.

12

It was one thing to say he had to learn about the skull, and quite another thing to do it.

Aside from taking it out of the closet again and looking at it—which he did on Saturday morning, turning it over and over—there didn't seem to be a way to know anything.

On closer examination in good light the skull proved to have all good teeth, no cavities. At least in the upper teeth —he did not have the lower jaw or teeth.

"Does that mean anything?" he said aloud in his room. It was early on Saturday morning and his mother was gone—off on some trip with Bill. They were growing closer and Brennan was happy for her—although it had happened many times before. Getting this close.

"It could mean he was young. . . ."

There it was again. *He*. Why did he think the skull

was male? No, why did he *know* the skull was male? Because he knew it, was certain of it.

And without any reason.

"All right." He set the skull on his desk, the eye sockets staring up at him. In a second it bothered him and he turned it sideways. "All right—so the teeth are good and that might mean it—*he*—was young. And if he was young, then the measurements could mean he was about my age."

Of course it's all guessing, he thought, leaning back. From the side he could see the damage done by the bullet. The entry hole in the forehead was a little over a half inch in diameter, and almost perfectly round. But a piece of bone as big as the palm of his hand was missing at the back, broken out in a rough oval.

God. How must that have been, he thought. How could that be? To have an explosion and then a bullet slam through your head that way and carry away the back of your skull and all the things you are, all the things you were or are or ever will be are gone then, blown away.

He shook his head. He was squinting, feeling the pain, and he tried to think of something else but could not. Instead he thought of the film he'd seen of Vietnam, an old film showing a man shooting another man in the temple on a street. It had been a television news film. He remembered the way the shock of the bullet had made the man squint.

He stood, turned away from the skull, looked out the window, and broke it then, broke the train of thought.

A week. I've had the skull a week and a day and I'm going crazy. What have I done?

He wrapped the skull up again and put it in the closet. I'll do it now, he thought, I'll call the police. . . .

But he didn't, couldn't. Instead he found himself putting his running shoes and shorts and a T-shirt on and heading out into the cool Saturday morning air.

He set an easy lope away from the house, not meaning to head in any direction but in a block he turned left and started the long road that went up and around the side of the mountain overlooking El Paso.

There was almost no traffic yet, no distractions, and he gave himself to running, did not think but increased the pace until he was driving up the mountain road, his legs pumping.

In moments the work made him sweat and he pulled his T-shirt off, still running, and rolled it and tied it around his forehead to keep the sweat from his eyes.

Running hard now, pushing himself, deep breaths, deep and down his legs knotting and bunching and taking him up the steep road, his shoes slamming on the road, no thoughts, a blank . . .

And the word *Homesley* came in.

Perfect, he thought. Homesley.

Maybe he'll know what to do.

13

John Homesley was a biology teacher in Cardiff School. Brennan had taken biology from him the year before and an almost-friendship had developed.

Well, Brennan thought, jogging down the street that led to Homesley's house, cooling from his run—it hadn't started that way.

Brennan had nearly flunked, had trouble in school, and Homesley had stopped him in front of the school one day as he was heading home at the end of the day.

He was an enormous man, tall and very heavy, bordering on fat but in a controlled way. Like a bear. He had rounded shoulders that somehow looked massively strong, with a heavy head of dark curly hair that always seemed a little long and a neatly trimmed beard filled with gray streaks.

"Did you know," he had said to Brennan, standing in the sun and grass in front of the school, "that the beetles are the most numerous species on earth?"

"What?" Brennan had been in a hurry and was slightly annoyed at the delay.

"Beetles. They're the most numerous species of life on earth. Do you suppose that means God made the earth for beetles?"

"I beg your pardon?" Brennan was confused. He had taken biology but Homesley hadn't said four words to him. As in most of his classes Brennan had taken a desk in the back of the room and spent much of his time trying to be ignored.

And now this teacher had stopped him on the school lawn and was talking to him.

"I said, beetles are the most numerous species on earth —so do you suppose that means God made the earth just for beetles? That beetles are God's favorite thing?" He stared down at Brennan, his eyes serious, but a faint smile at the corners of his mouth.

"I don't know," Brennan said, and thought, God, I sound dumb. Maybe I *am* dumb. "I guess so."

"Aren't you curious about them?"

"Beetles?"

"Yes. Don't you want to know about something that is the most numerous thing on earth?"

"Well . . . I don't know. I guess so. Yes. I guess I am curious about beetles." The sun was on him and he had to squint to look up at Homesley.

"Good. Let's find one."

"What?"

"Help me find a beetle. There's probably one within five feet of where we're standing."

And he put a pair of reading glasses on, which made his face look round and almost clownlike, dropped to his hands and knees and started looking through the grass, moving blades of grass sideways with his fingers.

Brennan stared at him. There were other kids going past and they stopped to watch.

"Come on." Homesley looked up. "Give me a hand. It's easier with two looking."

And still Brennan hesitated. Then he thought, oh, well, maybe it will help my grade and he dropped to his knees and started looking through the grass with Homesley.

"What are you looking for?" A tall kid in the eleventh grade stopped.

"Beetles," Brennan said.

"Beetles? You mean like bugs?"

"Yes," Brennan said, without looking up, wishing he could drop into a hole in the ground.

"They're the most numerous species on earth."

"Oh."

"Mr. Homesley wanted to find one and I'm helping him."

"Oh."

The boy had walked away shaking his head and Brennan kept his eyes down into the grass.

And he saw one.

A black bug, almost an inch long, a shiny black beetle. "I found one."

"What does it look like?"

"It's black and shiny and about an inch long." Brennan put his finger down and poked the beetle. It raised its hind end. He poked it again.

"It's probably a blister beetle. Don't touch it."

But it was too late. The beetle emitted a squirt of some kind of fluid from beneath its rear end, squirted it on Brennan's finger. Instantly there was a sharp burning sensation, a quick sting.

"Ow . . ."

"Exactly. They have a defense mechanism that's pretty effective. You'll hurt there for a while and might get a blister, but you'll be all right. Did you know that some beetles have a small turret gun down there and they can aim all around their body and hit with incredible accuracy?"

Brennan shook his hand. The spot on his finger hurt like a sting. "No. I didn't."

"Oh, yes, beetles are fascinating. A person could spend his whole life just studying them. Just beetles." He sat up, looking straight into Brennan's eyes, his face serious. "I can't believe you don't want to know things."

"But . . . well, I do."

"You don't seem to want to know biology."

"That's different . . . I'm just not good at it."

"Nonsense. You found the beetle, didn't you?"

"But that's not the same. . . ."

"But it is. Exactly. That's what studying biology—or anything else for that matter—is all about. Just finding things. Do me a favor, will you?"

By now there was a circle of kids watching and Brennan had never been so embarrassed in his life. He had spent most of his childhood being very shy and trying to not be noticed. And now Homesley had put him right in the middle of things.

"Every day bring me a different kind of beetle."

"What?"

"Bring me a new beetle each day and we'll learn about them together. . . ."

And a sort of friendship had developed. Brennan had done as Homesley had asked and brought a beetle, a different kind of beetle, each day and they would look it up, study the characteristics, talk about them.

And Brennan passed biology—Homesley had been as good as his word. But a strange thing had happened. Somehow working with Homesley had bled over into other parts of school. It wasn't that he enjoyed school—he wouldn't go that far.

But he studied. His habits changed and he studied almost by instinct; almost naturally.

Which just as naturally brought his grades up.

Which made his mother happy.

Which made him happy.

Which made it still easier to live, to study, to learn—all because of Homesley and his beetles.

But perhaps more important, Homesley had shown him other things as well—other than biology.

He had invited Brennan to his home one weekend, where he lived with a wife named Tricia who was almost as small as he was large.

He had taken Brennan into the basement, where Brennan had expected to find jars of bugs, or plants, or fetal pigs floating in preservatives.

Instead it looked like the interior of a space vehicle. Every corner, every wall was filled with electronic equipment.

"What . . ." Brennan stopped just inside the room.

"Music."

"Music?"

Homesley had nodded. "Trish and I love to listen to music. We can't watch television—it's too . . . slow for us—so we listen to music."

"You mean like rock?" Somehow he couldn't feature Homesley listening to wild music.

And Homesley had shaken his head.

"Mostly classical. I like Mahler, and Bach and Beethoven. Trish gets into opera."

There was an overstuffed couch in the middle of the room and at each end a floor lamp stood.

"You mean you just come down here and sit and listen to music?" he asked.

Homesley had nodded. "Sometimes we read—but usually we just listen."

"And all of this is just for music?"

Another nod. "Of course a lot of it is speakers. Would you like to try it?"

"I sure would. . . ."

"Sit on the couch, in the middle, and lean back. I'll play Mahler's Resurrection Symphony for you."

Brennan thought of asking for some Pink Floyd or Creedence Clearwater but decided Homesley probably didn't have them.

He sat.

And listened.

And it was more than just hearing the music. At first the strains of Mahler sounded soft to him, and he thought he would be bored—which was what usually happened when he listened to classical music.

But the system, the speakers made the sound so . . . so pure somehow, so rich and pure that the music went past just hearing, past listening—the music went into him.

He sat in stunned silence while the whole of Mahler's Resurrection Symphony, the whole of Mahler's music, the whole of Mahler's thinking went into his mind.

It was incredible.

When the music was finished he looked around the room expecting something, the whole world, to have changed. Homesley had left as the symphony began and he came in holding a can of soda.

"Like it?"

Brennan felt like whispering. "I didn't know, you know, didn't have any idea music could be that way. . . ."

And so Brennan learned about beetles and about music that year. Through the spring he went several times to the Homesleys' house and listened to music, talked about music, about biology, about nothing and everything and learned most about himself.

When summer arrived he went to work for Stoney and had not spent much time with Homesley. He had been working too hard.

But the friendship was still there and the feeling that Homesley was perhaps the only person who could help him with this skull business.

That's how he thought of it now.

This skull business.

No sleep, dreams he couldn't begin to understand, thoughts, voices through his brain that made no sense, his whole life upside down . . .

This skull business.

He jogged now to cool a bit coming off the run as he neared Homesley's house.

Homesley could help.

He hoped.

14

"A skull."

Homesley's voice was flat. He stared at Brennan.

Brennan nodded.

"With a bullet hole in it."

Another nod. "In the forehead."

"And you have it in your closet."

"Yes." Brennan sighed. "It seems pretty stupid, doesn't it?"

"Well." Homesley had been trimming a small green plant and he plucked a leaf from it, thinking. "Well, I don't know. Let me get all this straight. You found the skull in a canyon and brought it home without telling anybody?"

"Right."

"And you still have not called the police?"

"I can't."

"Brennan . . ."

"I mean it. I can't. I've started to several times, many times, but something stops me and I can't. I just can't. And I'm having all these weird dreams that I don't understand and things are happening to me. . . ."

He told Homesley all of it, from the camping trip to the dreams, and when he was finished Homesley leaned back and put his hands on the table that held the plants.

"My," he said. "My, my, my . . ."

Brennan felt drained, tired. "And so I thought I would come to you."

Homesley looked out the window, thinking. A fighter from Biggs Air Force Base roared over the house making a low-level run into the desert. Homesley's house was in a development not far from the runway and the fighters were frequent visitors. Strangely he didn't seem to mind the noise—although it was also true that it couldn't be heard in the soundproofed basement music room.

"I haven't seen the skull, of course," he said, thinking out loud. "So I can't say much about it. But from what you say it looks old."

Brennan nodded. "Very old."

Homesley smiled. "Well, not *very* old. Not prehistoric or even much over a hundred years old. It had to be after guns were here—there is, after all, the bullet hole, isn't there?"

Again Brennan nodded. "Yes. I'm sure that's what it is. It's neat and round in the front. . . ."

"And a large chunk carried away in the back."

"How did you know?"

Homesley smiled, a small sad smile. "I was a medic in Vietnam. I know something of bullet wounds."

"Oh."

"So—there wasn't any hair or tissue on the skull or around it, right?"

"Right. It was . . . clean."

"That means, I think, that we can assume the skull wasn't part of a recent murder and may not constitute evidence."

"You sound like a lawyer," Brennan said.

Homesley looked at Brennan, his eyes serious. "It may very well come to that—lawyers. If the skull *does* constitute evidence—you realize that, don't you?"

Brennan tried not to think of it but he had realized it.

"And that doesn't change your mind about contacting the authorities?"

Brennan hesitated. "It's not . . . my . . . mind. Sometimes I don't want to keep the skull but a thing takes over my thinking and I can't do it. Can't contact the police."

Homesley nodded. "I see. Or maybe I don't. But I think I know what you mean." He pushed the plant away and stood, walked back and forth in the kitchen, thinking and talking as he moved. Brennan almost smiled. Homesley looked exactly as he did when he was teaching—striding excitedly left and right in front of the class.

"Facts are almost nonexistent," he said, "so it's really difficult to come at this with logic."

Brennan nodded, but said nothing.

"So we take a few shots at probabilities here. Not very scientific but I think helpful. Jump in or correct me as we go, all right?"

Brennan nodded again.

"The skull was found in the canyons up by Orogrande." He named the small settlement on the highway—little more than a gas station—north of El Paso.

"Near Dog Canyon," Brennan added.

"So, Dog Canyon, if memory serves, is one of the last places where the Apaches and the soldiers fought."

"Bill said there were several battles there," Brennan said.

"Which means we can logically assume that there were probably fights in some of the other canyons—say where you found the skull."

"And that it's an Indian skull . . ." Brennan cut in.

Homesley held up his hand. "Not necessarily. We can guess that, surmise that, assume that, speculate that, but we cannot *know* that—not without examining the skull. Correction, without having an expert examine the skull. I'm not an expert at pathology but . . ."

"I thought you knew about these things," Brennan said. "I mean it's almost like a fossil, isn't it?"

"But," Homesley interrupted, and finished, "I do *know* an expert pathologist. No, Brennan, I do not know about these things." He smiled. "There are some things even I don't know. . . ."

Brennan frowned. "I'm a little worried about showing the skull to anybody else. Oh, God." He shrugged. "Listen to me. I sound like I've got something to hide. This is crazy—just crazy."

Homesley waited a moment, then sat back down at the table. "I know this is all troubling you, but if you're going to learn anything you're going to have to get help."

"I am. You."

"I'm not enough. But if it's any help this pathologist and I are good friends and I think I can promise you he won't tell anybody about the skull."

Brennan leaned back, waiting, and realized with a shock that he was waiting for a voice in his mind to tell him what to do. Little voices in my mind, he thought—oh good, I'm waiting for little voices in my mind. Oh great.

None came.

"All right," he said, standing. "I'll go get the skull and meet you at this man's place. . . ."

"Better yet," Homesley said, rising. "I'll call him and we'll take the car and drive down there. It might make it a little easier for you if I'm with you."

"What do you mean, easier. Where does he work?"

Homesley stopped with his hand on the door. "Where a lot of pathologists work—at the morgue." And he moved out the door.

Brennan hesitated only for a moment—thought, oh, the morgue, of course—and followed.

15

The morgue.

There was a room with metal tables and bright overhead lights and Brennan did not want to be in the room.

Everything in flat white, with large sinks along the wall and drains all over the place and the constant sound of running water and Brennan followed Homesley into the room and *really* did not want to be there.

On one table was the body of an old man. He was nude, lying flat on his back with his eyes open staring at the ceiling, his body all sunken and old and very, *very* dead.

Homesley's friend—who turned out to be named Tibbets—was tall and thin, wearing a smock spattered with blood and rubber gloves. He was leaning over the body holding a scalpel about to make a cut in the top of the dead man's stomach and Brennan almost lost his lunch.

He did not, *could* not turn away—but couldn't stand it either and he stopped.

Dead.

Tibbets looked up, saw them, and stood away from the corpse and Brennan thought, thank you, thank you, thank you . . .

"The boy with the skull," Tibbets said. "The big mystery." He and Homesley exchanged looks, then Tibbets came forward and took the tennis shoe box Brennan was carrying the skull in.

"Let's see . . ."

He took the box to a side table so that Brennan—gratefully—had to turn his back on the body.

Carefully, almost tenderly, Tibbets removed the skull. Brennan had wrapped it in a towel and Tibbets removed the towel gingerly. He put the skull on the table on a small foam pad and pulled down a large magnifying glass and light.

In the harsh white light against the metal table the skull looked very small and somehow sad, forlorn.

Tibbets must have felt it as well because he seemed to take extra care.

He first gave the skull a close visual examination through the magnifying lamp, turning it this way and that, over and around, humming songs from the sixties.

Brennan stood there and watched him and tried not to think of the body on the table in back of him.

Tibbets then took a hand-held magnifying glass and

examined the skull again still closer and much more slowly. He spent a long time looking at the bullet hole and brought another small, high-intensity lamp into play, held it almost in the hole.

He nodded, broke the humming for a second, then started again.

He turned the skull over and studied the teeth one at a time.

All of this took close to ten minutes and Brennan fidgeted from foot to foot.

Finally Tibbets took out two sets of calipers and made measurements. He measured overall the size—drawing a sketch of the skull on a pad with the part that had been blown away in dotted lines—and marking the measurements in small, precise numbers.

Then he measured the eye sockets with an inner set of calipers, again writing the numbers on the sketch, and when this was done he put the instruments down and turned to Brennan and Homesley.

"Well?" Homesley asked.

"Well, I don't think the skull is evidence, so you don't have to worry. But he did die violently."

"He?" Homesley rubbed his neck, stiff from looking over Tibbets's shoulder.

"Definitely. I'll tell you first what I know for certain, then the guesses, all right?"

Brennan and Homesley nodded.

Tibbets put the skull in the center of the foam pad and brought the light close down upon it.

"First—the wound. It's definitely a bullet wound. Judging by the size of it and the damage done to the back of the skull I would say it was a very large bore rifle. Say forty-five caliber or bigger."

He paused, thinking. "The wound was done from very close range. Probably inches—less than a foot."

"How can you tell?" Homesley asked.

"Because burned powder from the charge was actually driven into the bone of the skull. The scars are still there, the marks, and for that to happen the muzzle of the weapon has to nearly be touching the victim."

Victim, Brennan thought. Victim. He felt suddenly sick, weak. Imagine how it must have been, he thought. To have somebody hold the barrel of a gun to your head and pull the trigger. "Suicide," he whispered.

"What?" Tibbets asked. Brennan's voice was so quiet, neither of the men understood it.

"Could it have been suicide?" Brennan asked.

"Almost certainly not." Tibbets pointed to the hole in the skull and the damage done to the rear. "It was such a straight-line shot it just about couldn't have been self-inflicted —the angles are wrong."

"So this man was killed by a large rifle," Homesley put in. "Shot from very close range. . . ."

"Not a man," Tibbets said.

"But you said 'he.' "

Tibbets nodded. "I did. But not a man. He was a boy."

"A boy?"

Again Tibbets nodded. "It was an Indian boy—you can tell by the skull shape—about fourteen years old."

Take me, spirit. Take me up. It rolled into Brennan's mind and he reeled with it, almost fell over.

Homesley put out a hand to catch him. "Are you all right?"

"Don't know. Air. Need air."

Tibbets shook his head. "I forget sometimes where I work and how it gets other people. Let's go out to the lounge. . . ."

But, Brennan thought, weaving as they led him out of the room—but.

But no. I am not sick.

I am not sick.

I am not . . .

Me.

I am not me.

16

"Here, have some coffee." Tibbets held out a plastic cup.

Brennan shook his head. Slow, wobbly shakes that seemed to take forever. In the lounge, on the cheap couch, he thought—I'm here. "I don't drink coffee."

"Drink this anyway," Tibbets said. "It will take the shock away."

"A Murphy drip," Homesley said, looking at Tibbets, who nodded.

Brennan took a sip. It tasted like oil drained from a car. Sludge. "What's a Murphy drip?"

Homesley smiled that sad smile again. "We were both medics in Nam—that's where we met. A Murphy drip is something medics use. . . ."

"But what is it?"

Tibbets cut in. "You boil coffee down to a thick black solution. It's solid caffeine. When a man gets hit sometimes the shock—wound shock—will kill him even if the wound doesn't. You need a fast, raw stimulant and when we didn't have one we'd use the coffee."

"But what if they couldn't swallow?" Brennan was having trouble swallowing himself.

"No—you don't understand. We gave it anally. That was the only way to get it into the system fast enough to do any good. . . ."

Brennan looked at Homesley, who nodded to him. "That's right."

"But that's . . . that's awful."

"All of it—all of Nam was awful."

Tibbets looked over Brennan at something nobody else could see and nodded. "All of it . . ."

Homesley coughed. A soft sound. "Well—back to the present. Or the past. Finish telling us about the skull."

Brennan sat forward. Whether from the coffee or just getting out of the examining room he felt more alert.

"As I said—it was a boy of about fourteen. The teeth show little wear. Death was instantaneous and probably happened between 1860 and 1890—I would lean toward 1865 or so."

"You can tell all that from the skull?" Brennan was amazed.

Tibbets shook his head. "No. There is other knowledge involved, and I'm guessing a bit now." He thought for a moment. "It's a good bet that he was killed by soldiers—they used large-bore rifles and killed Indians. And the center of Indian—Apache—action in the canyon area was about 1860. It's just a guess. . . ."

He looked at Brennan. "You didn't say specifically where you found the skull—just in the canyons up by Alamogordo. Is it secret?"

Brennan shook his head. "Not from you. It was near Dog Canyon, a canyon south of it or maybe two. I don't know the name of it."

"All of those canyons are on National Forest land," Homesley said. "It probably goes without saying but you're not supposed to take things away."

Brennan didn't say anything, just looked at him.

"Well. In case you didn't know it . . ."

"I have to get back to work." Tibbets stood. "I have a suggestion—if you want some help."

Brennan nodded. "I sure do."

"You still know almost nothing about the skull—or rather about the boy who was the skull. You need more information."

"Where do I get it?"

"You could ask the Apaches up at White Mountain Reservation above Alamogordo. They might be able to help."

"I can't get up there," Brennan said. "I have to work. . . ."

"Or you can contact the other side—the army."

Homesley stood up. "Better yet—I have a friend who works in Denver in the Western Historical Archives. We'll write to him and ask him for any information he might be able to get us."

Tibbets stopped with his hand on the knob to the examining room door. "Ask him for all military reports, especially after-action reports for the time between 1855 and 1885." He nodded to them and went back to the body on the table before Brennan could thank him.

Homesley pointed to the exit. "Should we go do the letters?"

Brennan followed him out but stopped in the parking lot near the car. He held the box with the skull in front of him. Not like a skull, a bone fragment, but as a person. He felt a sense of urgency that he could not understand any more than he understood anything else that was happening to him.

Homesley looked at him over the top of the car. The sun was baking the metal and heat waves rose, made his face seem to move. "What's the matter?"

Brennan hesitated. "Well, I don't want to seem ungrateful but . . ."

"But what?"

"Do you think that if I paid for it we could call your friend in Denver? It will take days and days to send and wait for letters. . . ."

Homesley shrugged. "Sure. Why not? Let's go. . . ."

And Brennan had one fleeting thought as he got in the car.

Why didn't my mother meet this man and have me and he would be my father and . . .

He shook his head and sat back in the hot seat as they drove across El Paso.

17

It was easily the longest week in his life.

He worked hard with Stoney all week and called Homesley every night and when he learned nothing he ran, ran each night until he was tired and fell into bed and slept hammered into the pillow. There were no dreams.

"It takes some time to find all the stuff we asked for," Homesley said the first time he called. "Be patient."

It was easy to say, but very hard to do.

The skull was in his closet, in the box.

Waiting.

He couldn't shake the feeling that it waited—sat in the box and waited for . . . for . . .

He didn't know.

But the pressure was there, tremendous pressure, and just when it seemed he would explode Homesley called.

"It's here."

"I'll be right over."

"It's nine o'clock in the evening and there are several boxes. . . ."

"I don't care. I'll tell Mom I'm staying over—if it's all right, I mean."

"Of course it is." Homesley didn't hesitate. "We'll put you on the couch in the music room."

Brennan cleared it with his mother and was gone out the door, almost before she nodded.

He tried not to run fast the three miles or so to Homesley's—tried to set an easy pace. But he couldn't hold it. His legs, his mind, the skull took over and soon he was running at a dead lope and when he arrived at Homesley's he was sucking wind, his chest heaving.

"You must have run hard all the way," Homesley said, watching him try to catch his breath. "Are you all right?"

Brennan nodded. "Where is the stuff from the archives?" he gasped.

"In the music room."

There were seven boxes, each about a cubic foot in size.

Brennan picked up one of the boxes, put it down, picked up another. He didn't know where to start. He picked up another one and tore the top open.

Inside were photocopies of papers. Hundreds of sheets laid sideways in manila folders.

He pushed back the corner of one folder.

"Eighteen eighty-one," he said aloud. "He's dated each one. The work—the work he's done. It's incredible."

"Like I said." Homesley sighed. "He's a friend."

"He must be a very *good* friend."

"The best."

"Vietnam?"

Homesley nodded silently. "Were you all friends that way?"

Another nod. "Those of us who made it." He turned away and picked up another box. "He has them in order—see? There's a number on the side of each box. Also, there's a letter."

He opened a manila envelope and read aloud. "The boxes are in order by date. They start in 1855 and end in 1895. Inside each box the papers are also by date. There are army reports and letters and newspaper clippings. He goes on about some private stuff here, then he says he isn't sure what we want but he thinks there might be something for us in the box marked number three."

Brennan picked up the number three box and sat on the couch, opened it.

"Eighteen sixty-four," he read from the first folder.

He pulled out some papers that proved to be copies of old newspapers. Even as copies it was easy to tell that the original paper had been old. The newsprint was uneven and the headlines almost screamed.

" 'More raids by Apaches,' " Brennan read aloud.

" 'Livestock stolen, settlers frightened, army increases patrols.' "

"Got into their headlines, didn't they?" Homesley said. "They didn't need to write the story after that."

Brennan read the story. It said that there had been several raids in the area south and west of El Paso. That ". . . great damage has been caused." He finished the article. "It's so weird to read this—it's like it just happened. Yesterday or something."

Homesley set the boxes on the coffee table in front of the couch. "It's going to take days to go through all this—and we really don't even know what we're looking for, do we?"

Brennan shook his head. "Not really. But I'll know when I find it—I think."

He sat back on the couch with the box in his lap and started with the first folder. In moments he was lost in the papers.

He did not see Homesley leave, did not see Homesley come back with a plate of chocolate chip cookies and glass of milk, did not see Homesley put them on the table next to the boxes, did not see Homesley leave.

Brennan knew nothing but the papers, the papers and the world they showed.

18

Dust and heat. The newspapers spoke of patrols and raids and heat and dust. Long patrols the army made when the soldiers had to cut the veins on their horses and drink the blood; dust and sand so thick, they had to cover their mouths with bandannas just to breathe.

Brennan read the newspaper articles and stories first. They were all poorly written, redundant—they would often say the same thing several times.

"Twenty soldiers attacked a marauding band of Apaches in a running battle," a story said. "No soldiers were injured though several Apaches were seen to have been hit." And later in the same story: "A roaming patrol of twenty soldiers attacked a band of wild Apaches. After a brisk exchange of fire several Indians were hit though no soldiers received wounds."

He put the article aside without looking up, without seeing the cookies or milk, took another story.

This one spoke of the ranchers who had to take their families into El Paso because they feared an Indian attack.

Ranchers, Brennan thought—how could they ranch? The area around El Paso was hot, dead desert; sand dunes and mesquite and a few coyotes. How many ranchers could there be?

Again, nobody was hurt.

There was much writing of fear:

Indian Fright!

Ranchers Flee!

Citizens Terrorized!

But when he got into the articles he found that while livestock had been driven off, there had been no injuries caused.

Or very rarely.

Some prospectors had been attacked at their mine and one had died of wounds received in the battle—while four Indians were killed in the same fight.

So violent, Brennan thought, leaning back. He saw the cookies and milk and took some. Everything was so violent—white, red, color didn't seem to matter. Violence was the way of it—the engine that seemed to drive the West was violence.

All the articles were about Indian uprisings but on one copy, part of another article had been copied as well. It told of two soldiers who had hacked each other to death with knives in a cantina in Juárez. The author didn't seem shocked so much as amazed that they would kill each other at the same time.

Brennan finished the cookies and drank some milk. Then he started to read again.

He read some of the articles in the box and opened the other boxes and scanned the articles in those as well.

But he could not forget the comment about box number three and how it might have something for him. They had told Homesley's friend—he was named Ted Rainger—only that they desired information about the canyon area north of Fort Bliss and any actions that took place there between 1860 and 1890 or so.

The third box concerned El Paso and Fort Bliss and the canyons but it took some time to go through the other boxes to make sure.

He looked up one more time, saw that the wall clock said one in the morning, and knew he should try to sleep.

But he could not stop reading. He started into the general papers in the third box. They were mostly action reports with a fair number of letters.

The action reports were paperclipped to the order to which they applied.

"You will take your patrol to the vicinity of Hueco Flats where you will sweep north and south of the road from Carlsbad, engaging any hostiles encountered unless the enemy force is too large for engagement to be prudent. You will carry twenty rounds of ammunition per man and such rations for man and horse as deemed suitable for an eight-day patrol. [Signed] Captain John Bemis."

And clipped to the patrol order was the report of the patrol leader.

"We patrolled as per instructions the area of Hueco Flats with a fifteen-man unit all in standing good health, man and horses, and the patrol was without incident until noon of the fourth day when the patrol came under fire from small arms. No men were hit though one mount sustained a mortal wound. Fire was returned and a group of approximately twelve hostiles was seen to leave an arroyo a hundred yards distant and ride south at a good pace. No hits were registered and no pursuit was given. The rest of the patrol was without event, arriving back at Fort Bliss on 12 September 1864."

Brennan put down the folder holding the orders and visualized what he had just read.

The words were so dry, concise, but they meant so much. Horses running, men shooting at other men—men trying to kill other men. A horse hit going down, everybody yelling, the Indians riding out of the arroyo and the soldiers firing at them.

What was that like? To aim at another person with a gun to try to kill him. How could that be?

He saw the clock again.

Two in the morning now. I have to work tomorrow, he thought—no, today. I have to work today with Stoney. I should lie down and sleep. Right now. I should lie back and sleep and rest, but he could not.

Instead his hands took another folder out of the box and he began to read. And then another, and another, and he read until it was past four, until his eyes burned and his brain was so full of words, of reports and more reports, that he fairly reeled with them, read until he was on the edge of something like sleep. . . .

And then he found it.

19

It started with a patrol order much like the others.

"You will patrol the area north of Fort Bliss along the line of bluffs leading to Alamogordo, over to the base of the Organ Mountains and back to the originating point. The purpose of said patrol to maintain the public order and engage hostiles if engagement seems prudent."

Brennan's mind perked. The "bluffs leading to Alamogordo" was the canyon area. This was the first patrol order in that area. But it was the after-action report that held him.

"Reporting on patrol 21 Oct. 1864. We worked due north of Fort Bliss without incident. On the second night, approximately thirty miles north of the originating point, our pickets reported hearing horses in the night but no contact was made with the suspected hostiles.

"We rode north two more days with no contact.

While making the sweep across to the Organ Mountains the patrol intersected a group of hostiles driving a herd of approximately 100 horses.

"Chase was given with the troop in skirmish formation. The main body of hostiles abandoned the horse herd and escaped in the direction of White Sands.

"Two hostiles bringing up the drag on the herd were seen to break away and ride toward the canyons in the line of bluffs to the east.

"Four troopers—O'Bannion, Rourke, Daneley, and Doolan—were dispatched and gave chase. One of the hostiles was killed in a running fight and the second was cornered in a canyon south of Dog Canyon. The troopers reported that after a brisk exchange of fire the second hostile was killed. Due to the remaining length of the patrol the bodies were not returned."

"Chaaach!" Brennan's breath exploded. He had been holding it without realizing it. A canyon south of Dog Canyon.

A coldness was in him now, a deep feeling of certainty. There was no real reason for it, for the sureness, but he was absolutely positive the skull belonged to the "hostile" the soldiers had chased into the canyon.

The canyon.

He had run then, run from them up into the canyon. One boy, his age, his own age, one boy running and four men after him.

Names. He knew the names of the men. All Irish.

Probably big men, soldiers, blue uniforms, hot, stinking of sweat, chasing him, chasing me into the canyon up beneath the rock. . . .

He shook his head, rubbed his eyes. Not me, *him*—not me. The two kept mixing in his thinking. The soldiers chased him, not me. I am sitting here reading about it—not running into the canyon away from them, trying to hide. . . .

The clock again. Five-thirty now. There were no windows in the basement but outside the sun would be up. It was morning.

He took out another packet of papers from the same folder.

Letters. The copies were all the same size but it was easy to see that they had all been written on different sizes of paper, at different times and by different people.

He riffled through them, wondering if he had time to read them. He almost put them aside but remembered Homesley's friend Rainger had worked hard to organize everything. The letters must mean something or they wouldn't be there.

He rubbed his eyes again. Two hours before he had to go to work.

He read the first letter.

20

"To the commanding officers of Fort Bliss.

"Hoping to find you in good health I take up my pen in a matter most urgent.

"Here at the Quaker school and home we meet often with both young and old Apaches. As you know we are in God's work to help these people and bring them to grace.

"These are violent times and require more open methods than in more civilized eras and so I shall come to the point.

"Thursday last I was approached by an older Indian woman from the Horse Mesa band of Apaches. I had not seen any member of that band in some time and could only hope that they were off hunting.

"But she informed me that such was not the case and her band had made a raid to steal horses in Mexico.

"On the raid two young boys were killed. One boy was named Magpie and your soldiers apparently shot him while riding and his body was recovered.

"The second boy was seen to enter a canyon with some troopers chasing him.

"This second boy was named Coyote Runs and the old woman was either his mother or his maternal grandmother. These relationships are sometimes hard to understand but I think she was the boy's mother.

"She said since they cannot find a body would I use my good offices to ask the army if they truly killed her son and if so, would they tell her (me) where the body is so she may retrieve it and give it a proper burial.

"I remain your respected servant,

"(Mrs.) Amelia Gebhart."

Coyote Runs, Brennan thought—his name, the skull's name, was, is, Coyote Runs.

"Coyote Runs." He said it aloud. A boy my age named Coyote Runs ran into the canyons with four soldiers after him. Firing at him.

I fear.

I fear for you now, Coyote Runs, I fear for you.

He put the letter down, picked up the next, which proved to be an answer.

"Dear Mrs. Gebhart.

"Having received yours of last month I take leave to reply.

"Regarding your request concerning the action of a patrol in the area of the canyons adjacent to Dog Canyon, I fear I have not much help for you.

"I have spoken personally with the officer in charge of the patrol and subsequently the four troopers who were personally involved in the aforementioned action.

"Their report assures me that while they did dispatch two hostiles in a running fight the second Indian was also killed in the running fight. Due to the heat and the remaining time of the patrol there was no attempt to recover the bodies of either man.

"It is to be expected that in the heat and with the number of coyotes this year nature may have taken its course, not to put too delicate a turn to it.

"I am sorry to have been of so little help.

"Your respectful servant,

"Col. John McIntire (Comd.)"

He's lying, Brennan thought. Whether he knows it or not, he's not telling her the truth. I found the skull up in the canyon, way at the back. With a hole in the forehead. There is no way he could have moved after being hit that way.

What was it Tibbets had said?

Oh yeah, instantaneous death.

Coyote Runs had come into the canyon. They had chased him.

He must have been hurt. Maybe his horse fell or something. Maybe they wounded him.

He had run from them but they had cornered him under the rock and shot him in the head.

Wait.

More—there was something more. There, yes, there it was—Tibbets had said the muzzle of the weapon had been held right against the boy's head.

They hadn't killed him.

They had executed him.

Run to earth like an animal, run to a hole like some frightened animal and they had found him and leaned down and put the barrel of the rifle to his forehead, put it right there and pulled the trigger.

"Oh."

Brennan blinked, put a finger to his forehead. There.

There.

And Coyote Runs and all the things he would ever be ended then, in that instant.

Brennan put the papers on the table and leaned back. Tears moved slowly down his cheeks.

To know him—to know his name and how he must have died and lived.

No. That was wrong.

He knew his name now and how he died but he knew nothing of how Coyote Runs must have lived.

The door opened slowly and Homesley was standing there. He was holding a steaming cup.

"It's time to get up," he said. "I brought you some herbal tea . . ." His voice trailed off. "You've been up all night."

Brennan wiped his cheeks with the back of his hand and took the tea. He held it with both hands and smelled the strong herb steam. "Thank you."

"You found out something." Homesley said it not as a question but as a statement of fact. "Something about the skull."

Brennan sipped the tea. "His name was Coyote Runs and four soldiers executed him."

Homesley sighed. "Things never change."

"I want to know more," Brennan said. "I want to know all about him, how he lived, how it was for him to be."

Homesley nodded. "I understand."

"There is a thing I must do—I know what it is now."

Homesley said nothing, waited.

"A thing I am supposed to do."

Still Homesley waited.

"I have to take him back."

"Back where?"

"Back to where he is supposed to be."

"Where is that?"

"I'm not sure, exactly. Somewhere up in the canyons. He'll . . . tell me when it's right."

"He'll *tell* you?"

"I know how that sounds—crazy. And maybe I am. But it has to be—this has to be done."

"I understand."

And Brennan could see that he meant it. "I'll go get the skull—go get Coyote Runs. Do you suppose you could take me up to the canyons and drop me for a couple of days? I'll tell Mother I'm going camping. . . ."

"Sure. No sweat. I'll take a sleeping bag and go with you."

"We'll take our time and see the country up there. . . ."

And Brennan was only half right. He would certainly see the country—but it wouldn't be leisurely, and it wouldn't be with anybody.

21

Bill's car was parked out front and Brennan could sense something as soon as he entered the house—in the air. A tightening tension.

"Is that you, Brennan?" His mother heard the door open and close. "Come into the kitchen, please."

Her voice was flat, but not angry.

Brennan went into the kitchen.

The scene seemed frozen.

Bill sat on one side of the table and his mother sat on the other. Bill's eyes were wide, curious; his mother had some of the same look to her face mixed with concern.

Between them, on the table, sitting on a towel was the skull.

The skull.

Ahh, he thought. Ahh—there it is.

"Do you have an explanation for this—this hunk of bone?"

Yes, Mother, I do, he thought. His name is Coyote Runs and he died when an American soldier put the muzzle of a rifle to his forehead and blew his brains out. But his mouth didn't say the words. Instead his lips opened and he said, "I found it when we were camping."

"And brought it home without telling me?"

Brennan shrugged. "There were all those kids with us and I didn't want to make a mess. . . ." It sounded lame and he knew it, let it trail off.

"That was National Forest land," Bill said, his voice prim. "You're not supposed to remove things from National Forest land. . . ."

"Coyote Runs."

"What?" Bill asked.

"Not 'things,' not 'hunk of bone.' His name was Coyote Runs. He was a fourteen- or fifteen-year-old Apache boy killed by the army."

"How could you know all that?" Bill asked.

"Research. We . . . I've been doing some research." He thought it best not to mention Homesley.

"See?" His mother put her hand on Bill's arm. "I told you we didn't have to call the police."

"You were going to call the police?" Brennan asked.

"We already did," his mother said. "They should be here any minute."

"Ahh, Mom . . ."

And he could see it all then—everything. He could see the police coming in and the questions they would ask and what they would have to do—take the skull. They would have to confiscate the skull and it would go back to the National Forest, back to the government and they would put it on display in a museum somewhere, stick it in a glass case with a little plaque saying what it was and he could see it.

See it all.

And as he saw that he also knew what he had to do; what he had to do immediately.

He must take Coyote Runs back. He must follow the dreams, the instructions in the dreams. Now it was all there, all clear; he knew what would happen and knew what he must do.

He must take Coyote Runs home.

He grabbed the skull from the table, made for the back door.

"Brennan." His mother's voice stopped him. "What are you doing?"

"Mother . . ." There was nothing he could say to make it right for her; no words to explain what he had to do. "Mother—I have to leave for a few days."

"Why?" She stood. "What are you talking about?"

"He . . . needs me to do something."

Outside he heard a car pull up in the driveway. The police were arriving. Great, he thought—just great. Now I'm running from the police. I haven't done anything wrong and I'm running from the police.

"Who?" his mother asked. "*Who* needs you to do something?"

Brennan raised the skull. "Him."

"The *skull*?"

"Coyote Runs. He needs me to—well, just needs me."

"But Brennan . . ."

"I'll be back in two or three days. I have to do this. . . ."

"But . . ."

"Mother. Please."

"But . . ."

And she either nodded or he thought she nodded. Either way it no longer mattered. Time was gone. Brennan moved through the back door and was gone.

22

The darkness seemed complete, thick, black, close around him, and he took it as a friend. There was a time when he was afraid of darkness and it surprised him faintly that he had changed.

Changed so much.

He had set an easy pace north. For all of the day, through the morning and into the afternoon he jogged lightly through the city and out past the suburbs and Fort Bliss and into the desert.

Fifty, sixty miles.

It was between fifty and sixty miles to the canyon country where he had found the skull. He remembered that from the camping trip.

He had never run or walked any distance like sixty miles—across a desert—yet he did not doubt he could do it.

The skull—Coyote Runs—would help him. He did not know how, or why, or where he was going but he knew he was not alone.

Take me, spirit . . .

Fort Bliss extended some miles north of El Paso—missile-firing and artillery-impact ranges overlapping. Signs along the road warned of unexploded artillery shells. For that reason Brennan stayed along the road until he thought he was well past the last signs, then he turned into the desert.

He had stopped at a convenience store and purchased a small knapsack and a couple of granola bars and a quart of bottled water. He put the skull and water in the pack and adjusted it to ride loosely on his back.

When darkness had first caught him he stopped and lay against the side of a sand dune to rest and stretch his legs. He had rested several times during the day just for ten or fifteen minutes each time to keep his legs from cramping. He had also drunk most of the water—amazed how fast a quart seemed to go.

He had stopped using the water in midafternoon, when the sun was hottest, and as he lay in the dark to rest he felt his cracked lips with his fingers. They were bloody. He had taken a small mouthful of water then, enough to wet his lips, and gone into a fitful kind of half sleep.

When he opened his eyes it was pitch dark, dark that wrapped him, and he thought for a moment how nice a fire would be. There was a chill in the air and dampness that came down on him.

I won't be able to see to move, he thought.

But as soon as the thought came he began to make things out. Clumps of mesquite on sand dunes, openings between the dunes appeared first as shadows, then took on solid shape and meaning.

There was enough light from the stars and a sliver of a new moon for him to see where he was going.

Off in the night he heard the faint sound of cars on the highway as they passed.

He stretched, bent at the waist and pulled at the backs of his leg muscles until they almost snapped.

Then he put the pack on his back and started off, moving easily between the dunes. He found the Big Dipper then the North Star and he set off to follow it.

And he knew things then without knowing how it was he could know them.

It was this way in the old times.

The voice—more a felt sound, an echo in his mind, than an open sound. It was his own voice, in his mind, but not his mind.

It was this way in the old times that when a man ran in the desert at night he would push a stick ahead through the sand for the small snakes.

Brennan found a length of mesquite, broke it off about four feet long. One end had a slight curve and he held this end to the ground so that it slid along the sand.

He started moving again, a half run, half shuffle, the stick sliding along in front of him.

A good trick, my friend, he thought. I had not thought of snakes. My friend. Yes. My friend.

It was this way in the old times that when a man was thirsty he would find the stalk of the plant with long quills and eat of it and find sweet water.

The plant with long quills? What could that be?

Oh. The yucca. That must be what he was talking about.

He turned out a sidewinder that had been half buried in the sand. He flicked it off to the side with the stick and it buzzed angrily before looping off to the side in the darkness.

And I knew that, he thought. I knew he wouldn't bother me.

I *knew* it.

Off to the edge of a dune he saw a yucca plant, the spiny base looking like a porcupine hunkered in the shadows. He went to the plant and squinted, trying to see better in the darkness.

Sticking straight up from the center of the plant was a stalk about three feet long. He pushed at it, touched it, and found it to feel heavy, soft.

He worked it back and forth and broke it off even with the base down in the spines.

With a little more effort he broke the stalk in two and put one broken end to his lips.

The end fairly dripped with liquid. He tasted it and found the fluid to be a sweet-tasting water—almost like watermelon.

He bit and chewed on the stalk, eating it a few inches at a time, swallowing the sweet water gratefully before spitting out the pulp and taking another bite.

Thank you.

It was this way in the old times. . . .

Thank you. He ate the whole stalk as he trotted, broke and ate another and moved through the night in the easy trot, refreshed by the sweet juice from the yucca and the coolness of the desert night.

Well before dawn he sensed light, felt the temperature drop a bit more and stopped to rest.

No. It is not time to stop.

But . . .

They are too near. We must not stop.

What . . .

Go now. Keep going now.

I am tired. I am so tired, he thought. But his legs kept driving, pushing, pounding.

Another snake—this time not a sidewinder but a larger rattler—was pushed into a warning rattle by the stick.

Brennan moved easily a step to the right as it struck at the stick and kept moving. This, he thought, from a boy who was afraid of spiders and snakes.

He had changed in some basic way. He was still Brennan but more, much more, so that he was part of the night and part of the desert and part of the sky and part of the snake and knew these things, knew all about them and did not fear them.

Into the light from dawn he moved; until the buttes and canyons rose above him to the right, he moved; until he could see boulders and lines he moved. He could not, not move.

Stop.

The thought, word, feeling, cut through the fatigue, the movement, and he stopped.

It was here. Here where the pony soldiers found us and chased us and brought Magpie down. His chest went out before him.

Here.

Brennan felt an enormous sense of sadness, a deep remorse. An old friend died here. A close friend.

And here we ran, and here, my horse running well but not fast enough and there by the cut across the land they hit my horse, hit my leg and I was down.

Brennan felt it, knew it and fell to his knees.

Here—it was here that it all happened.

He heard noises, heard sounds and thought for a moment that time had warped and that he would turn to see the soldiers riding at him.

But the sound was real.

Voices.

He heard voices out ahead—faint. Near the canyon mouth.

He held his breath and listened.

Men. It was the sound of men talking. He could not

understand the words but heard them as a low sound, a familiar low sound.

Then he heard one voice he knew.

His mother's.

His mother had come with men to look for him.

To stop him.

"Mother," he said half aloud. "Mother is here with some men."

He moved closer to the voices, keeping sand dunes and low bush between them for cover and staying low.

When he was fifty yards away he stopped again, lowered to his stomach, and—dragging the pack—made his way closer until he lay not thirty feet from his mother.

She was dressed in jeans and tennis shoes and was wearing one of Brennan's old jackets. There were two sheriff's deputies standing near her, and Bill, and off to the side a few paces, Homesley.

"He hasn't been acting normally," Brennan's mother was saying. There was worry in her voice. "Ever since the camping trip when he found the skull."

The deputies said nothing for a moment. One of them nodded to her and the other looked up into the canyons and coughed politely into his hand. "You think he ran here from El Paso in one day and a night. . . ." The tone of his voice was clearly skeptical.

"Maybe not yet." His mother stopped the deputy. "I don't think he could be here. But we have to stop him *before* he gets into the canyons. I'm worried about what may happen if he goes up in there alone."

"He'll be fine." Homesley had been silent all this while, standing off by himself. But he moved toward the others now. "There's nothing whatever wrong with Brennan. He just has something he must do."

"What?" His mother's voice was sharp, hard-edged.

"I'm not sure. Something with the skull. Maybe take it back. I don't know for certain."

"It appears to me," Bill cut in, "that if you hadn't helped him with all this, none of this would be happening."

"No." His mother held up her hand. "None of that. Brennan has learned a lot from Mr. Homesley—I won't hear people throwing blame around. There's nobody to blame about anything—I'm just worried about Brennan."

Good for you, Mother, Brennan said to himself silently. I could end it here, just stand up and end it. What am I doing anyway? This is crazy, wacko.

But his legs did not move him. He did not stand.

Instead he slid backward on his stomach until a dune with mesquite all along the top hid him. It would be a simple matter to walk around them and head up into the canyon.

But a new sound stopped him.

Engine. Truck engine. He worked forward on his elbows, dragging the pack with the skull up to the top of the dune where he lay in the mesquite and could see all below where the desert road ended in front of the canyon.

Bill's van was there, and the sheriff's car with the engine running.

But now a carryall wagon pulled up. On the side were written the words MOUNTAIN RESCUE.

It came to a halt in a swirl of sand and two men jumped out. They were dressed in climbing clothes.

"We came as fast as we could," one of them said to the deputies. "How long has he been lost?"

"Who are these people?" Brennan's mother asked.

"Rescue is always called in when somebody is lost." The deputy shrugged. "It's standard procedure. Always."

"But he's not lost," his mother said. "At least I don't think so. . . ."

"Still."

"How long has it been?" the rescue man repeated. "Dehydration may be setting in while we stand here talking."

Not likely, Brennan thought, remembering the yucca stalk. Then he felt like a snot—here all these people just cared about him, were worried about him. And he was treating them like they were enemies. Well, he thought—enough of this. He wouldn't let his mother worry.

He stood up.

23

It was like a scene from a bad movie. The boy gets lost, Brennan thought, and they find him in the nick of time and everybody lives happily. . . .

"There he is!" one of the rescue men yelled. "Right there!"

Brennan was only fifty feet away from them and looked at his mother.

"I didn't want any of this," he said. "Not for you to worry—none of it."

"I know," she said. "But . . ."

"I have to do this," he said.

Homesley stepped forward. "I brought them when your mother called me. She was worried."

"I know. It doesn't matter." Brennan stopped. The deputies and the men from the rescue unit were spreading out.

"You ran here in one day?" Homesley asked.

"And a night. Yes. He taught me what to do."

"Remarkable."

"Just a minute." Brennan held up his hand. "What are you doing?" The men had moved apart, slowly, seemingly without reason, but always apart so that they now faced him but spread away from each other.

"Nothing," one of the rescue men answered, smiling. "Why?"

"Don't come after me. . . ."

"You have to realize that you can't just take off up into the canyon."

"Why not?" Brennan tensed his legs. He closed his hand tightly on the knapsack holding the skull. They were going to come for him. My God, they were going to make a move on him.

"We have your mother's call that she was worried about your . . . status. You might be depressed, upset—we can't allow you to go up there alone. You might get into trouble. We can't allow that."

"I take back my complaint." Brennan's mother brushed hair out of her eyes. The morning sun was directly in her eyes and she had to squint. It made her look like a picture he'd seen of her when she was young, squinting into the camera, and Brennan smiled.

"It doesn't matter," the rescue man said. "He could still get hurt—he can't be allowed to go up there alone. . . ."

"Run, Brennan," his mother yelled suddenly. "Go now!"

It surprised Brennan so that for an instant he did not move. But only for an instant. The two rescue men jumped suddenly toward him and the deputies moved to the side to get between Brennan and the canyon mouth.

Brennan moved without thinking. He virtually fell backward off the dune through the mesquite and brush.

His legs windmilled beneath him and he somehow landed on his feet. The two rescue men came loping around both ends of the dune, arms outstretched to grab him.

He had one moment when he almost smiled—wondering what they were supposed to be "rescuing," then he felt their hands on his arms.

He rolled left and right quickly, slipped from their grip, and was away.

Now, he thought—or thought he thought. *Now away. Up into the canyon.* The voice was there, in him, of him, around him.

Take me, spirit . . .

He scrambled up the trail, heard the men coming behind. They were experienced at climbing, in good shape, but Brennan moved faster. His feet found the old trail and he ran lightly, exulted in his freedom. I can fly, he thought, and made the mistake of looking back.

Looking back once to see if the two men were gaining on him. His right foot came down on a rounded stone and his ankle turned. He felt/heard it pop.

An almost-click and pain roared up his leg and he went down on his side.

No.

It was this way before. They came and I was hurt in one leg. Do not stop. Do not stop. Do not stop. You must not stop. Ever.

They had him. He felt their hands grabbing him.

"No!"

One leg, he thought. I have one leg.

On his hands and the one good leg he rose, looked into, directly into the face of one of the rescue team.

"Listen," the man said. "Take it easy and listen to me. We don't want to hurt you . . ."

But Brennan didn't hear him, didn't see him, didn't see the rescue man.

His eyes were wide, he felt the heat and dust and heard his own breathing as wind whistling and echoing in some empty place.

And he saw a cavalry soldier.

Saw the blue wool of his uniform, the skin of his neck over the collar of his tunic, the stains from his sweat; smelled the horse in his clothing, saw the stubble on his chin.

All real.

"No!" he screamed again, and kicked backward with his one good leg and broke loose once more.

Then up.

Then up into the rocks and the heights of the canyon.
Never to stop.

Up to safety.

Up.

24

They would not give up.

Two of them.

They must have left their horses at the bottom of the canyon.

Good.

Even with one bad leg no bluebelly on foot could catch him. No two bluebellies could catch him.

I am, he thought.

I am Apache!

He fought the pain down, made his mind think of other things. His mother. *See his mother bending down to pick up wood. See his mother how she works. See how she builds the fire and moves the pot onto the flame. See . . .*

Brennan shook his head.

Can't, he thought. Can't remember things that never happened to me. How can that be?

He looked back. The men had fallen still farther behind. His leg was on fire but he was moving well. Sweat poured down his face and he wiped it away with the back of his wrist.

Where?

Where am I going?

He recognized parts of the canyon from the camping trip. Even limping he was making good time.

But where?

The canyon rose above him, around him. But the canyon would come to an end. He remembered that as well.

Back up there it ended, he thought, looking to the cliffs that marked the rear of the canyon. And then what?

They'd have him.

Take me, spirit.

There it was again. The voice. The whispers.

But I can't, Brennan thought. I can't go where I can't go, can't go, can't go . . .

Still he climbed the trail until he was moving across the flat, grassy meadow; moving now without purpose because he was sure he would not be able to get out of the cliffs that ringed the back of the canyon.

The sun cooked now, directly over the canyon. Even running—more jogging with a limp than running—he could not help but see the beauty of the canyon. The cliffs had looked red in the morning light but were now showing bars of gold and purple in the rock strata.

He saw the boulder now up ahead, where he had

found the skull, and for a moment he thought that was it—where he was supposed to go.

He thought that perhaps the skull wanted only that—to return to the place by the boulder to be with the rest of the skeleton, which must still be there.

But he knew he was wrong—something felt wrong about it.

What?

Not the boulder . . .

It was here—just so it was here. I ran, even with one leg I ran to reach the sky, reach the blue sky and freedom; ran to reach the medicine place but it was here, here that I ended. . . .

Brennan drew closer to the boulder. He had not known how it had been, and did not know it now. But he could feel something of it.

How the soldiers must have followed him—as the rescue men followed him now. Brennan hesitated as he neared the overhanging boulder, almost stopped and as he did the men gained on him, grew close enough so he could hear their feet.

No. Go on, go on and up and on to the place of medicine, to the ancient ones. . . .

There was great urgency in it, in the voice, in the feeling and he started to run again, past the boulder.

The sudden jump into a run twisted his sore leg and the limp grew temporarily worse and it slowed him. He could hear the men gaining again, their feet slamming into the dirt.

Ahead there was a ledge and a drop ten feet down into the streambed.

Brennan went off at a full trot and tumbled head over heels into the streambed in a cloud of dirt and loose rocks near the spring and pool of water. It looked cool, and inviting, but the streambed moved still farther up the canyon and he knew he had to continue.

The fall had knocked the wind out of him but he scrambled to his feet and started moving again, always up.

It looked worse all the time. The sides of the creek-bed grew steeper and higher and he could see no way to get out.

"You have taken me the wrong way," he said aloud between breaths.

But he kept moving. The streambed drew narrower still and began to wind sharply left and right and tighter and Brennan slowed, finally, thought it would end here.

Watch for a sign, the line of fire.

Ahh, he thought—sure. Watch for a sign. Fire? The line of fire?

And there it was, up and to the left.

In the wall of the streambed on the left side, about even with his forehead, there were several lines gouged in the rock. They were jagged and very old, faded and nearly gone and had he not been watching for them he would have missed them.

But he saw them and even as blurred as they were he knew they represented a bolt of lightning.

The line of fire.

Next to the blurred lightning bolt there was a fissure in the rock. It was narrow, went straight up the rock wall, and Brennan would have passed it, the way it lay back and partially out of sight, had he not been looking for the sign.

He looked back in the fissure—large crack in the cliff face, really—and saw two handholds just above his head, one on each side.

Up.

One word. He put his hands in the handholds and pulled himself up.

And there were two more.

And up, and two more, and still more and in twenty feet the fissure—which had been just wide enough for his body—began to widen and he saw that the handholds went on up the split in the rock all the way to the top of the canyon wall.

Three, four hundred feet.

Up.

Did he do this? Brennan felt the ridges with his fingers as he climbed, felt the edges where a tool had carved and chipped. They were weathered, old—very old. Ancient. Coyote Runs had not done this—the handholds had been there for hundreds, perhaps thousands of years.

But he had used them. Coyote Runs had used them. His hands and feet in the same place, pulling up.

Up.

Brennan kept climbing and when he was nearly fifty

feet up he paused and heard the two men come running up the rocky streambed. He held his breath—though they could never have heard him breathing—and the men ran past the fissure without pausing. Brennan smiled.

And it is so that I would have been free, been safe, if I had beaten them to the medicine steps. . . .

Brennan did some mental calculating. Maybe a quarter of a mile to the end of the canyon, perhaps a bit more.

Then they would start back. Ten, fifteen minutes. They would have to move slowly, looking for tracks, some sign of him in the rock streambed.

Maybe another ten, fifteen minutes. It was hard to estimate. Perhaps a total of twenty or thirty minutes until they found the fissure.

Then they would start up. They were expert climbers and would have no trouble.

Twenty minutes would have to be enough for whatever it was he was supposed to do.

He held for a moment, catching his breath and resting his legs, then he climbed again, looking up the fissure. The handholds were closer together now, made for a shorter body and it made the climbing faster, easier.

Above was the sky.

Blue sky. Not a cloud showed above in the narrow slot of the fissure. He climbed with his back to one side of the opening and his feet on the other, using the hand- and footholds on each side, and concentrated so hard on the climb that he didn't seem to notice how high he was getting until he nearly fell.

His hand slipped from one of the cutouts in his haste and had his foot not caught a hole on the other side and jammed him he would have dropped.

It was then, near the top, that he first looked outside the fissure and saw below him.

The canyon fell away beneath him in a drop that seemed to go forever. He could see out, past the mouth of the canyon, see the police and rescue cars there—they looked like toys—see the small specks that he knew to be his mother and the rest of them and his heart nearly froze.

He had never been this high, never climbed this way, and fear held him and he thought he would not be able to go on, to climb the short distance to the top.

Take me, spirit . . .

And his hand moved. He moved his eyes away from the view out of the fissure and looked only up. His hand found a new hole, then his feet and he skitched his back up, then a new handhold, then a foot, and up, and up and slowly he kept moving, kept climbing until he was at the top, until he could pull himself out on the flat rock of the top of the bluff where he lay for a moment, breathing deeply.

Then he stood.

And saw the world.

That was the only way he could think of it—he saw the world. The desert lay below him. He stood on a flat almost-table of rock that jutted out, formed one side of the canyon, and below and away lay all the world he knew. To his left, a haze in the south, lay El Paso, and across the great basin of desert he saw the Organ Mountains, gray and jagged, like broken teeth, and to his right, white and brilliant, lay the white sands, shining in the midday sun so bright it seemed the desert had been washed and bleached and painted.

And he saw this, Brennan thought.

He saw this same thing.

Coyote Runs was here, came up here, saw all this when he was still alive.

When he'd started the climb up the fissure he'd looped the handles of the knapsack around his neck so the pack with the skull would ride in front and be out of the way. He took it off now, let it hang in his hand.

To his right and perhaps twenty feet out on the flat rock was a squarish rock that seemed out of place. It was about a foot high and roughly two feet square, rounded by wind and weather at the edges, but Brennan's eyes were drawn to it.

The medicine place . . .

Of course, Brennan thought. Of course it is. . . .

He walked to the rock. In a small depression near it the limestone of the butte was a blackened color where there had obviously been fires.

Here it was, Brennan thought, here it was that he came to sit and learn things and know things. This was how he lived. How he was. Until they killed him.

The ancient ones are here, are always here in the medicine place. . . .

I understand, Brennan thought. He looked across the desert, stood by the rock and the small fire place and looked across the desert and understood why Coyote Runs had tried to get here, tried to be here, had brought him here finally in all the craziness of the run and the climb and the skull.

It was how he was, Brennan thought. It was the way of Coyote Runs to be here, to come here, to have his spirit set free here to go back to the ancient ones and he had been stopped, held back all these years, all this time.

Leave me, spirit . . .

Brennan nodded, smiled. He took the skull carefully from the knapsack and put it on the square rock, set it gently so the eyes were looking out at the desert and sky and moved away, stood away.

Be free, he thought, be the sun and wind and desert and stone and plant and the dust, be free and all the things there are to be.

He whispered, "Be free . . ."

And for a time there was nothing, for minutes or seconds or years there was nothing but the skull sitting on the rock and the world below.

Then a warm wind, a short, warm wind brushed his cheek and caught some dust from the rock around the skull and took the dust in a swirling column up, up over the bluff, over the canyon, over Brennan and into the sky and gone, gone where the dream of the eagle had flown, gone for all of time, gone and the skull was just that, a skull, a bone, nothing more, empty, as Brennan was empty.

He sighed, tears on his cheeks, missing something he did not understand, feeling that he had lost a friend somehow, and turned to meet the rescue men.